REVEALING

THE BURIED PAST

GEOPHYSICS FOR ARCHAEOLOGISTS

This book is dedicated to Dr Arnold Aspinall, former Head of Department at the School of Archaeological Sciences, Bradford University, and Chief Editor of *Archaeological Prospection*.

REVEALING THE BURIED PAST

GEOPHYSICS FOR ARCHAEOLOGISTS

CHRIS GAFFNEY & JOHN GATER

TEMPUS

First published 2003, reprinted 2004

Tempus Publishing Ltd
The Mill, Brimscombe Port
Stroud, Gloucestershire GL5 2QG

British Library Cataloguing in Publication Data.
A catalogue record for this book is available from the British Library.

ISBN 0 7524 2556 0

Typesetting and origination by Tempus Publishing.
Printed in Great Britain by Midway Colour Print, Wiltshire

CONTENTS

ACKNOWLEDGEMENTS

This book would not have appeared without the dedication of the staff at GSB Prospection. At the outset therefore we wish to thank all our employees, both past and present, who have surveyed over 1,700 sites on our behalf. We are very grateful for all the hard work undertaken. Their professionalism and enthusiasm for archaeological geophysics is commendable, if not remarkable, and this book contains many insights gleaned from them all. It is perhaps unfair to single out individuals; however, particular thanks are due to Dr Susan Ovenden and Dr David Weston who, in addition to carrying out many of the surveys, have provided valuable information on the sections on geophysical techniques and soils. Mention should also be made of Dan Shiel who has been responsible during the past 10 years for overseeing many of the projects included here. The writing of the book has been a long haul and would not have been accomplished without our CAD Manager and long-time field surveyor, Claire Stephens. Claire has performed wonders in mastering the majority of the final figures reproduced in this book. Her tireless efforts during the 10 weeks prior to completion of this book are gratefully acknowledged. Thanks Claire – without your efforts the preparation of this book would have been even more traumatic!

Dr Arnold Aspinall, who inspired a generation of archaeological geophysicists, introduced both of us to the subject and has been extremely supportive over the years. He kindly read and re-read the drafts of the book and his advice and comments were invaluable; however, any errors and omissions are our sole responsibility. In a similar way Dr Roger Walker must be thanked, not only for supplying images, but for his thoughts and advice, though again he is not responsible for the end product. Early versions of the text were read by Sue Gaffney and Dr Susan Ovenden and we thank them for correcting many errors. Any that remain are our own.

All of our clients who have given permission to produce versions of data previously only seen in unpublished reports are acknowledged: David Wilkinson, Oxford Archaeology; Rick Turner, CADW; Ed Dennison, Anthony Walker and Partners/BHWB and the Highways Agency; Robin Jackson and the 'Four Parishes Archaeology Group'; Drew Shotliff, Albion Archaeology; RCHME; Professor Mike Fulford, Reading University; Nick Card and Julie Gibson, Orkney Archaeological Trust; Jane Downes, Orkney College; Patrick Ashmore, Historic Scotland; Steve Dockrill, Bradford University; John Williams, Kent County Council; John Samuels, John Samuels Archaeological Consultants; Peter Rose,

Cornwall Archaeological Unit; Jim Symonds, ARCUS; Bill Britnell, The Clwyd-Powys Archaeological Trust; Royston Clark, CPM; Malcolm Atkin, Worcestershire Historic Environment and Archaeology Service; Simon Buteux, Vince Gaffney and Roger White, BUFAU; Roger Martlew, Leeds University; Charly French, Cambridge University; Cambridge Archaeology Unit; Gwilym Hughes, Dyfed Archaeological Trust; Nick Russil and Peter Brabham, Terradat.

Thanks are due to Tim Taylor (Series Producer) and Phillip Clarke (Executive Producer) of *Time Team* for involving us in the programmes over the past 11 years and ensuring 'geophys' has received a high profile. We also thank Professor Mick Aston, who was in part a catalyst for this book, for his support throughout the television series and on other projects where we have collaborated.

People who have freely given their own data are gratefully thanked. In particular colleagues at English Heritage (Andrew David, Paul Linford, Neil Linford, Louise Martin and Andy Payne) have helped us over the years and have delved into their archives on our behalf. Colleagues abroad who have supplied data or images include: Wolfgang Neubauer and Alois Eder-Hinterleitner; Bruce Bevan; Ken Kvamme; Larry Conyers; Dean Goodman; Juerg Leckebusch and Alet Kattenberg for organising the Dutch castle data from RAAP at short notice.

Finally we apologise to anyone we have forgotten; there must be many.

FOREWORD

by MICK ASTON

This is an important book. Chris Gaffney and John Gater are two of the foremost archaeological geophysicists in the world today. They have written here a superb introduction to the background and modern practice of geophysics on archaeological sites.

I first met Chris and John during the Shapwick Project in Somerset in 1989 when I saw for the first time the miraculous appearance of a site – in that case the buried medieval church – on a computer screen in the back of their jeep. They say in this book that they are not magicians, but to me on that day in a Somerset field they were. Not long after that they became the geophysicists for the *Time Team* programmes on Channel 4 television and have appeared in over 100 programmes in the last 10 years. Time and again they have shown the clear and essential role of geophysics as a modern archaeological technique.

This is the first archaeological geophysics book for a generation and follows the seminal books by Martin Aitken and Tony Clark. In it is revealed the fascinating story of how geophysics became a powerful tool for archaeological surveying, the stages by which the instruments were developed, and the close progress alongside computer technology. There are many case studies of exemplary surveys.

It is therefore a pleasure to welcome and recommend this book which will prove interesting and useful to archaeologists, and to those members of the public who may have seen the technique used on television and would like to know more about it.

PREFACE

When we embarked upon the writing of this book it was not without some feeling of trepidation. Over the years it seems a long time has been spent in the field collecting data, attending conferences and writing short articles. When we were approached by Peter Kemmis Betty from Tempus it was neither the first offer nor the first time that we had seriously given consideration to such a publication. Somehow on this occasion the timing was right. Whether it was due to increasing age, loss of hair or simply a failing ability to remember facts and figures, we decided that the time had come to have our say on archaeological geophysics.

Today the techniques of archaeological geophysics have probably never been so widely used or the discipline had such a high public profile. The arrival of commercially-led archaeology has dramatically affected the number of surveys that are carried out, a figure thought to be now in excess of 450 per year in Britain alone. The successful exposure of the techniques to a wide audience within the UK beyond the archaeological community is due almost totally to the highly popular archaeology programme *Time Team*, transmitted on Channel 4 television (www.channel4.com/history/timeteam). The authors have been involved with the series since its inception in 1993 and the first transmission in 1994. We have now completed over 100 shoots, the series regularly attracts viewing figures of over 3 million, and it has been estimated that over 20 million people, a third of the entire British population, are familiar with archaeology thanks to the programme.

Geophysics is an integral part of *Time Team* and a whole range of differing techniques have been used. In fact, a new television term – *geophys* or *geofizz* – has been coined to refer to the techniques. So common has 'geophys' become, that Professor Darvill of Bournemouth University recently advocated, albeit perhaps tongue in cheek, that the word should be a candidate for inclusion in the Oxford English Dictionary. It is now recognised by viewers, developers, and, dare it be said, even the more sceptical of archaeologists, that geophysics has an extremely important role to play in all aspects of archaeology (Cave-Penny 1995). In this book, many of the sites investigated by *Time Team* are used as examples of what can be achieved with geophysical techniques. We will also use samples from our own database of some 1,700+ surveys carried out during the past seventeen years by GSB Prospection. With so many surveys under our belt we decided that we had enough material to write a book that would be valuable to up-and-coming geophysicists and archaeologists alike.

Reading this book will not turn you into an archaeological geophysicist; it may convince you that you need a geophysical investigation on your site or integrated into your project. It may even make you want to become a geophysicist! It was our opinion that the lack of a modern introductory textbook has revealed a gap in knowledge that was widening into a chasm. More than any other objective we hope that this book will shorten, if not bridge, the divide between those who undertake geophysical surveys and those who benefit from the results.

Thornton, West Yorkshire
Summer 2003

SECTION 1

SCIENCE AND METHODOLOGY IN ARCHAEOLOGICAL GEOPHYSICS

In these five chapters we introduce the science behind the main techniques used in archaeological geophysics. Coupled with this is information concerning how these techniques are employed and how the data should be reported. These chapters contain information that is essential in understanding the case studies in Section 2.

1

A HISTORY OF ARCHAEOLOGICAL GEOPHYSICS

What is archaeological geophysics and why should we use it?

World Heritage monuments like Stonehenge and Hadrian's Wall provide a visible and tangible glimpse of man's past. Yet these monuments are but the tip of an archaeological iceberg since far more exists below the ground than survives above. Antiquarians and archaeologists have long recognised this fact and have been striving to locate these buried remains for centuries. In the past, this search may have been in order to plunder such sites, more recently to excavate them, while preservation is now firmly on the agenda. A variety of techniques have assisted archaeologists in this work; for example, the study of earthworks and historic maps, the mapping of artefact scatters, and, since the early twentieth century, the use of aerial photography. In the past 50 years or so, archaeologists have turned for help to more specialised methods, in particular geophysical techniques, techniques that have long been used in the study of the Earth and for the exploration of mineral resources.

While most people are aware of the destruction of archaeological remains by development, the rapid and widespread erosion of sites by agricultural processes has largely gone unnoticed outside of archaeological communities. The speed of this devastation, from all agencies, is frightening, and the need for fast non-invasive survey has increased dramatically. Archaeological geophysics is able to help carry out this work. In this book we will chart its brief history, look at the techniques used, the science involved and the results that can be achieved. We will also consider the future of geophysics in revealing the buried past.

Before commencing with a brief history of the subject we should define what we mean by 'archaeological geophysics'. There is in fact no dictionary definition; for the purposes of this book, we suggest the following:

> The examination of the Earth's physical properties using non-invasive ground survey techniques to reveal buried archaeological features, sites and landscapes.

From the start, we should state that in this book we are not concerned with geochemical methods, remote sensing (be it airborne methods or the non-destructive survey of standing buildings and structures) or any aspect of marine or underwater geophysics.

The emphasis will be on projects in Britain, where it is probable that more archaeological geophysics takes place than anywhere else in the world. But it should be remembered that we do not operate in splendid isolation; the exchange of information and ideas with colleagues abroad is a fundamental part of our work. It is hoped that this fact will be recognised by the occasional divergence into exotic case studies or through the numerous references included in this book.

In Britain today there are fewer than 100 full-time qualified archaeological geophysicists working in the field. They are employed by English Heritage (a government body), independent (commercial) groups, university departments and archaeology units. Furthermore, a large number of amateur groups exist that have their own equipment, such is the popularity and availability of the instruments.

So how have we arrived here today and how have archaeologists used geophysical techniques?

Archaeologists have always pondered the best place to locate exploratory trenches on their site. Indeed, one of the most famous archaeologists of all, Lieutenant-General Pitt-Rivers, catalogued his own non-invasive testing in the last decade of the nineteenth century. He reported in his *Excavations in Cranborne Chase* that he prospected and found the position of ditches by slamming the flat part of a pick on the ground and listening to the change in tone. This was simply a search using a basic geophysical technique termed 'bowsing' and one that inquisitive, even desperate, excavators have used the world over.

Although Pitt-Rivers' test was exemplary and well documented, it was only a first step in the quest for locating archaeology using physical attributes of the soil. For most people, a survey in 1946 at Dorchester on Thames is the starting point for this subdiscipline (Atkinson 1953). However, two recent papers in the March 2000 edition of the archaeological geophysicist's flagship journal *Archaeological Prospection* reveal a less Anglo-centric view of history.

The first accurately documented survey was not in England but in Williamsberg, Virginia, USA (Bevan 2000). The survey was undertaken by Mark Malamphy, a Canadian working for a Swedish geophysicist called Hans Lundberg. Lundberg was a trailblazer with a considerable number of 'firsts' in applied geophysics. The bibliographic details in Bevan's systematic re-evaluation of the work at Williamsberg show that Lundberg was an early worker in electromagnetic and inductive techniques. In 1921 he carried out the first aeromagnetic survey from a balloon; he may have undertaken the first electromagnetic survey from the air in the 1940s; and he developed a vertical component magnetometer for use in a helicopter in 1946. Bevan believes that had it not been for a car accident, it is likely that Lundberg himself would have undertaken the work at Williamsberg. History records, however, that Malamphy, an exploration geophysicist who worked widely in North and South America, was the person who conducted the first systematic geophysical survey for archaeological purposes.

The archaeological problem that Malamphy set out to resolve was identified by Marie Bauer Hall, who had found an earlier buried church adjacent to Bruton Parish Church at Williamsberg. She also believed that a stone vault, approximately 3m square, was buried about 3m beneath the western end of the earlier church. It was she who decided to engage a geophysical company to investigate the site. The technique used was an 'equipotential' survey. Bevan describes this method below:

> Two parallel lines of grounded electrodes frame the rectangular area to be explored. An electrical generator sends a current through the earth between the pair of wires; this current has an audio frequency. The pattern of the voltage at the surface is then mapped. This is done with a detector that comprises an audio amplifier and a set of earphones. Two wires connect the amplifier to a pair of metal electrodes that are driven into the soil. Wherever the two electrodes contact the earth where the voltage is different, the audio signal is heard in the earphones; if the voltage is the same, there is no sound. These lines of zero relative voltage are traced on the earth.
>
> If the earth is uniform, the electrical current flows directly from one of the grounded current electrodes to the other. Lines of constant voltage would then form simple and rather straight lines, which are parallel to the two lines of grounded electrodes. If something unusual is buried in the earth, the flow of the current will change, and the voltages also will be different. The lines of constant voltage will then be warped into irregular patterns where the unusual object is buried.
>
> (Bevan 2000, 53-4)

This survey was undertaken on 3 November 1938, a key date in the discipline of archaeological geophysics. Malamphy found an area of high resistance in the western part of the church and concluded that it could be a result of the vault, but he was evidently cautious and noted that the anomaly may also be the product of natural changes (**1**). Hall decided to dig but failed to reveal any cause for the geophysical anomaly (Hall 1974). Almost 50 years later Bevan, utilising the most advanced techniques at his disposal, searched for the same vault and found a high resistivity anomaly in the same place as Malamphy had done. The subsequent excavation found no vault and the anomaly was believed to be due to differential leaching of small fossil shells in the sandy sediments at a depth of about 4m (Bevan 2000, 56).

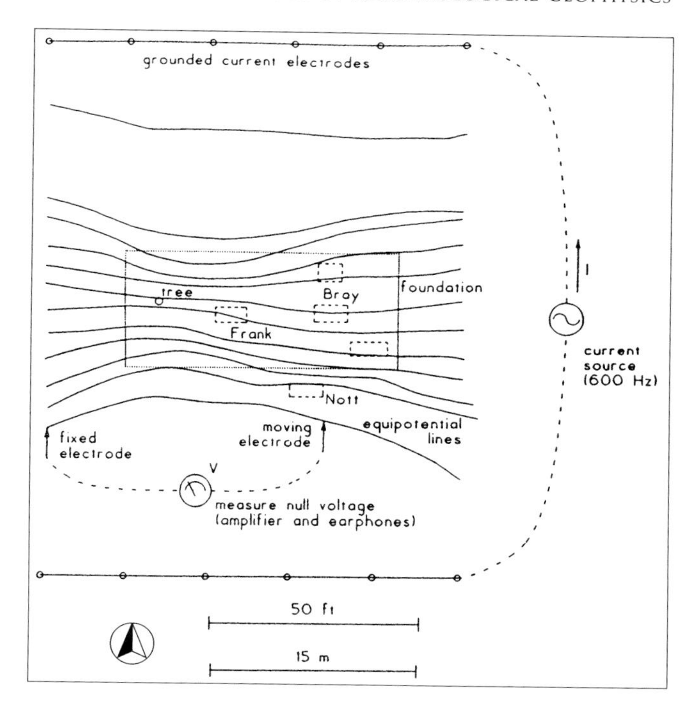

1 *A schematic layout of the first authenticated geophysical survey for archaeological purposes. The concentration of equipotential lines towards the left-hand side of the survey area suggests a buried high resistance feature. Although the surveyor, Mark Malamphy, interpreted the results as indicating a partially filled vault, more recent work has suggested that the anomaly was produced by variation in the subsoil.* This diagram was redrawn by Bruce Bevan (Bevan, 2000)

This highlights some important lessons that are still valid today. Firstly, archaeological anomalies are part of a wide range of causative bodies that are often difficult to differentiate. Secondly, even if a survey is professionally executed, there is still no certainty that a successful archaeological result will be achieved. A geophysicist is not a magician; the interpretation should be logical and based on the theory of the technique. Thirdly, if features are not discovered in a survey it is possible that even the 'best' documented or topographic evidence can be wrong.

A second, less known, geophysical survey was documented by Albert Hesse – himself a pioneer in the theory and application of geophysical techniques in archaeology, as well as the author of the first textbook on archaeological geophysics (Hesse 1966). In his article, Hesse (2000) provided a robust defence of the pedigree of researchers who did not publish in English. In particular he charted the work on metal detectors, chiefly 'electromagnetic' devices, that found not only a bronze dagger at the Bleiche-Arbon site (von Bandi 1945) but also reported that materials such as tile and brick caused significant signals on these devices (Laming 1952). The latter was an observation of Count Robert du Mesnil du Buisson, a French Oriental archaeologist who excavated mainly in Syria. Hesse also emphasised the prophetic nature of Mesnil du Buisson's 1934 book *La Technique Des Fouilles Archeologiques*, where according to Hesse's translation Mesnil du Buisson clearly identified the potential use in archaeology of seismic, gravity, resistance and magnetic techniques:

> It is too early to say how these methods will provide help for archaeology. Buried buildings made of big blocks, mainly of basalt, or hypogea composed of several chambers could probably be sometimes surveyed with such methods. At very extensive sites, where either stone buildings among cities made of bricks (Bactres) or royal tombs in a mountain (Kings Valley) are looked for, application of these methods will have to be checked.
>
> (Mesnil du Buisson 1934, translated in Hesse 2000)

The importance of these insights is the innovative statements, especially when one considers that geophysics was itself a fledgling subject during the early part of the twentieth century. Hesse argued that the consideration of techniques so recently discovered probably had a link with the great mining geophysicist Conrad Schlumberger, whose pioneering work with electrical survey was conducted just 43km from the Mesnil du Buisson's family chateau in Normandy. Schlumberger's name lives on in field geophysics as evidenced by the resistance configuration named after him (see chapter 2).

While the Williamsberg survey may have been the earliest systematic survey, the first survey to have an impact on archaeology was without doubt the resistance work by Richard Atkinson at Dorchester-on-Thames in 1946. Clark (1996) gives an excellent account of Atkinson's work which was published by Laming (1952) in French and greatly influenced the second edition of Atkinson's *Field Archaeology* book (1953). Atkinson hired a simple commercial device, a Megger Earth Tester, and developed a wafer switch to change between a series of five probes to produce four 'live' contacts, thus devising an efficient survey method based on the so-called 'Wenner' array. The survey results were plotted and the presence of ditches and pits was suggested. This time geophysics was seen to get it right! The site comprised relatively moist ditches cut into dry natural gravel. In short it was a perfect example to prove the value of the technique.

Atkinson's success tempted many archaeologists in Europe to try resistance survey as a way to find the best place to excavate. A knock-on effect was to concentrate the minds of many individuals to assess the additional techniques highlighted by Mesnil du Buisson 20 to 30 years earlier. The next revolution in prospecting occurred when the archaeologist Graham Webster was excavating sites during the widening of the A1 road near the Roman town of Durobrivae. Following a lecture by John Belshe on magnetic dating, Webster wondered if such techniques could be transferred into the field for detecting *in situ* kilns. Belshe had proved that this was possible in an early example of experimental archaeology when he had 'detected' the magnetic response associated with a reconstruction of a Romano-British kiln. In search of a practical answer to his problem, Webster contacted Martin Aitken and Edward Hall at the Research Laboratory for Archaeology and the History of Art at Oxford University. They were interested in a new proton magnetometer that had been developed for field use, the results of which were published that year (Waters and Francis 1958). The main problem was that they had only eight weeks to design and build a robust instrument that could find buried archaeology. They met the challenge and had an instrument in the field by March 1958.

Reviewing the work, Martin Aitken sketched out the practical problems they had encountered. Once the survey was underway, however,

> The first anomaly detected was due to an iron water pipe and subsequently we confirmed that iron bedsteads were magnetic too . . . . It was not until the last day that we found a kiln, beneath an anomaly 10 foot across and having a strength of 100 gamma at its centre. After so many days of fruitless searching the archaeologist benignly allowed us to dig our own investigatory trench (perhaps he was sceptical) and I shall not forget the excitement of uncovering, at a depth of 40 inches, the red rim of a kiln wall. It was the first of the many hundreds, or perhaps thousands, of kilns detected by various of us around the world . . . . Although we did not know just how successful an archaeological tool the proton magnetometer was going to be we celebrated appropriately that evening at the Haycock Hotel, Wansford.
>
> (Aitken 1986, 16)

In the same article Aitken expressed his initial surprise at the discovery of filled-in rubbish pits. As a scientist he was dismayed at he thought that the responses associated with the pits simply increased the noise around his targets. He was soon convinced of the merits of pits from an archaeological perspective and the true value of magnetometry, not as a kiln detector but to map slight but meaningful anthropogenic anomalies.

In the 1958 spring edition of the first ever volume of *Archaeometry* Martin Aitken wrote:

> the value of magnetic surveying in locating whatever archaeological remains are buried in an unknown region can be gauged from the experience that in one field 20 random trial holes all proved blank whereas 4 holes dug on magnetic anomalies revealed archaeological features in 2 cases, a geological feature in the third, and a horseshoe in the fourth.
>
> (Aitkin 1958, 25)

In the world of archaeological science the ramifications of this work were enormous. Within a decade prospecting was a research area that young scientists wanted to work in. As early as 1961/2 the pages of *Archaeometry* (a journal dedicated to science in archaeology) reflected the widespread interest with applications reported in Switzerland, USA (in particular the work of Elizabeth Ralph

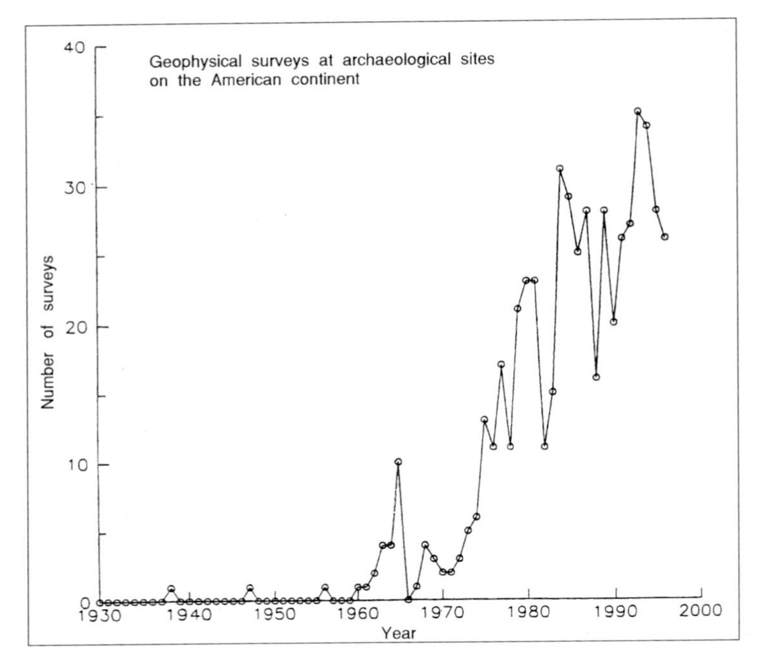

2 *This graph shows the increase in the use of archaeological geophysics on the American continent. Elizabeth Ralph was a great pioneer during the 1960s and her work was built upon by notable exponents such as John Weymouth and Bruce Bevan.* The graph was compiled by Bevan (Bevan, 2000)

at MASCA), France, Germany and Italy. The last three were to produce fine research establishments that effectively inspired many of the theoretical and applied areas of archaeological prospection. In France the CRNS centre at Garchy produced many significant papers by Albert Hesse and Alain Tabbagh; in Germany Irwin Scollar, who was based at the Rheinisches Landesmuseum at Bonn, contributed widely to the subject; in Italy Richard Linington, along with Carlo Lerici at the Lerici Foundation produced groundbreaking work in the southern part of Europe.

The Lerici Foundation also contributed to the dissemination of knowledge by producing, in 1966, the first volume of *Prospezioni Archeologiche*, a journal dedicated to the use of geophysical techniques in archaeological prospecting. Looking back through that journal one can see a maturing discipline with papers on every imaginable technique and an ever increasing sophistication in data display and processing. All the key researchers of the period found time to publish in *Prospezioni Archeologiche* – Aitken, Aspinall, Hesse, Linington, Scollar, Tabbagh, Tite, and Weymouth. The journal continued to be the main publisher of academic articles until 1986. *Prospezioni Archeologiche* had documented the scope of prospecting as well as the apparent end of an era. The importance of the journal is that any would-be researcher should check its list of articles prior to embarking on a project – in nearly every case someone has already published a note of some interest in *Prospezioni Archeologiche*. Of course many of the articles from that period are inconclusive, largely as a result of the technological limitations of the equipment rather than any deficiency in the technique.

As far as Britain was concerned, the establishment in 1967 of a Geophysics Section within the Ancient Monuments Laboratory (AM Lab), itself part of the Inspectorate of Ancient Monuments, was a major step forward in the development of archaeological geophysics. The first person employed was Tony Clark; subsequently the section has grown to a permanent staff of around six within English Heritage. As a government body its role has been diverse, ranging from testing new instruments and methodologies, to carrying out surveys on ancient monuments and more recently advising on the use of archaeological geophysics.

The AM Lab was involved in the development of the fluxgate gradiometer for archaeology (see chapters 2 and 3), an instrument that first appeared in the mid-1960s (Alldred 1964). This essentially paved the way forward for routine magnetic surveying in Britain. In the late 1960s, Philpot (1973) developed a smaller instrument and later conceived the idea of connecting it to a continuous chart recorder. Working in conjunction with Philpot, the AM Lab refined the system (Clark and Haddon-Reece 1973) and for the following 15 years this method dominated large-scale fluxgate surveys. The procedure adopted is described in detail in Clark (1996); the beauty of the set-up was that for the first time large area surveys could be carried out without the need to write down readings. The disadvantage was that the only archive of the data was the X-Y traces produced on the chart recorder.

Looking back at the late 1970s and early 1980s, there appears to have been an air of retrenchment in academic research in Archaeological Science, especially in Britain. Although some of this may have been due to the prevailing political mood in the universities (looking for different, novel avenues of research), in some cases it was more a shift in interests. The Oxford group that had been so strong during the sixties moved on from prospecting to analysis and dating – perhaps they believed there were too few fundamental academic challenges left. The deaths of Lerici and Linington saw the effective collapse of the Italian group. For a researcher during this period it was difficult to find published articles because of the lack of a specialist journal. For some, however, this was a relative period of development. In America there was a steady build up of surveys in this period (Bevan 2000, figure 1-2) and authors such as John Weymouth and Bruce Bevan were prolific.

During this period, however, a significant new agenda was drawn up in British universities emphasising the teaching of archaeological geophysics as opposed to fundamental research into the subject. This started with the creation in 1971 of a postgraduate course titled Scientific Methods in Archaeology, within the Postgraduate School of Physics at Bradford University. Run by Arnold Aspinall, geophysical methods were taught for the first time in a British University as part of an archaeology degree. The postgraduate course was followed by an undergraduate degree in Archaeological Sciences, offered from 1975. The Bradford School in effect took over from Oxford and the AM Lab by publishing research papers and articles in a wide variety of fields and has remained at the forefront of research in archaeological geophysics. The Bradphys resistance meter was developed, along with the innovative 'Twin-Probe' configuration. Software was also refined and the first successful attempts were made to couple instruments with the new generation of portable computers (Kelly *et al.* 1984).

Although there had been many published articles describing novel instruments and how to undertake geophysics, for the first time practitioners were considering the role of geophysics in archaeology. One article that summarised the 'state of the art' at the beginning of the 1980s was by Peter Fischer (1980). In that publication Fischer evaluated 'resistivity', 'electro-magnetic detectors', a 'soil conductivity meter' and a 'subsurface interface radar' as part of the site investigation. It is clear on re-reading the article that Fisher struggled with the limitations of the equipment and the problems of processing and display. Who wouldn't, given such a broad scheme? It seems that there was a need to prove each and every technique and while the audience hoped that the project would be a success, they very much doubted that it would. Geophysical techniques appeared too slow, too temperamental and too costly to fit into the increasingly large-scale landscape studies of interest to archaeologists.

In the mid- to late 1970s, changes were taking place outside the universities that would have a major effect on the future of the subject as a whole. In particular, an appointment in a government-owned company was in many ways a precursor for what was to become widespread practice in the commercial world ten years hence. British Gas, who were about to embark on a major construction programme of pipelines servicing the recently discovered gas fields in the North Sea, decided to appoint their own archaeologist, Phil Catherall. He worked first on the construction of a major pipeline across several archaeologically rich areas in southern England and, in collaboration with the AM Lab, used geophysics to investigate a number of sensitive points along the route (Catherall *et al.* 1986). Recognising the benefits of the techniques in archaeological practice, Catherall persuaded British Gas to appoint their own full-time archaeological geophysicist and in 1979 one of the authors (JG) took up that post. Overall, this was a new era in applied archaeological geophysics and as we have charted above, in universities there was a swing away from pure research to the application of geophysical techniques in archaeological fieldwork. With the first graduates from Bradford University starting to filter into archaeological posts at the start of the 1980s, a number of people were keen to, and capable of, undertaking routine survey.

Several major technical advances were made that would aid the collection of fast and reliable data. In 1984 Roger Walker, a former PhD student from Bradford, with research interests in archaeological geophysics and electronics, set up Geoscan Research. He produced rugged, lightweight and waterproof instruments that proved to be extremely reliable and probably made 'commercial' work a viable option. At this point the Philpot, Littlemore and Plessey fluxgate instruments became virtually redundant. A discussion of the instrumentation is provided in chapter 3.

It was apparent that archaeology was changing more towards landscape type issues and some thought that geophysics could blend in to the new approaches. Heron and Gaffney (1987) first categorised the use of prospection and archaeological research into 'intra-site analysis', 'site delimitation', 'off-site prospection' and 'site location'. Their article clearly argued for a more versatile use of detection devices and one that could engage in the location of elements within the landscape that had been changed by human intervention, but were not necessarily the walls and ditches of settlement. So convinced of the role that prospecting devices should play in archaeology they stated:

> we feel that practical archaeogeophysics deserves to be part of the theoretical debate in archaeology. It is hoped that archaeogeophysics integrated within a wider archaeological framework is seen as a legitimate avenue of practical and analytical research. Although it is naïve to suggest the cessation of excavation, prospection embodies the currently economically dictated themes of 'minimal impact' and 'cost-effective' archaeology and allows the gathering of information outside of excavation boundaries. There is more to archaeogeophysics than locating habitation structures for excavation. Archaeologists do not follow walls these days, why should archaeogeophysicists?
>
> (Heron and Gaffney 1987, p.78)

Although the authors attempted to compartmentalise the use of geophysics too rigidly, it was a radical and positive agenda. At the end of the 1980s a conference titled 'Geoprospection in the Archaeological Landscape' was held and many of the same themes were expanded upon. In the introduction to the proceedings volume the editor states:

> The published literature provided little cause to believe that geophysical techniques were being used in any way that could not be described as glorified wall-following.
>
> (Spoerry 1992, p.2)

In reality, however, although there was an academic agenda that argued for implementation of geophysical techniques in a wider archaeological arena, the impetus for a greater role came from elsewhere. While the inevitable trickle down of technology into the discipline, in the form of reliable equipment and 'instantaneous' digital data, created a platform from which to work, this in itself cannot be regarded as the reason for the subsequent explosion in activity. It has been argued elsewhere (Gaffney and Gater 1993) that the information required by archaeologists changed during this period – the rapid evaluation of large tracts of land became the norm, and where traditional avenues of investigation were weak, geophysical techniques were strong. In particular the development boom of the late 1980s/90s and the general absorption of archaeology into the environmental assessment of large-scale developments, provided great incentive for those attempting to establish geophysical techniques within the archaeologist's methodologies.

It has been estimated that in 1980 about 60 surveys were undertaken in England, Scotland, Wales and Northern Ireland. The two main providers were the Ancient Monument Laboratory and British Gas, who both carried out 30 to 40 surveys. By 1990 this figure had increased to about 250 per year (**3**). The significant differences between 1985 and 1990 result from the change in the people who were commissioning work. The major group was now under the banner of 'developers'. They were not asking for a survey for any altruistic interest in an archaeological site, but rather were concerned about the ramifications of potential archaeology within the context of a proposed development.

Spoerry has described the changes as follows:

> by 1991 something of a minor revolution appears to have occurred in the use of geophysics as part of the planning process. More and more planning authorities and archaeological units appear to be waking up to the value of properly executed, efficient geophysical surveys, carried out by experienced staff, using modern, reliable hardware and producing interpretations that stand up to excavation . . . . Perhaps the business acumen of individuals outside the profession has allowed such concepts to be accepted more easily than amongst the rather conservative ranks of the digging community. None of us are under any illusions that developers would readily fund all necessary archaeology if allowed a free hand, but, where planning constraints are imposed or threatened, it is undoubtedly in the developers' interests to carry out an evaluation that is as reliable as possible, as well as one that is relatively cheap. If that means using geophysics, then geophysics will be used.
>
> (Spoerry 1992, 5)

Although most countries have seen an increase in the use of geophysical techniques, archaeologists outside of Britain do not use them with such regularity. There are many reasons for this; some are a result of technological and software changes, while others are more fundamental. As Spoerry has pointed out, this rise in the number of surveys has come about without any real legislation or national policy. To date, the only national guideline that has given credence to non-destructive techniques is Planning Policy Guideline 16 (PPG16) (DoE 1990). In Britain the use of geophysical techniques is often said to be underpinned by PPG16. A crucial aspect of the guideline is that anything that is asked of the developer is 'fair reasonable and practicable'. Reasonable is usually interpreted financially and given that non-invasive work is cheaper than excavation, geophysical techniques will always be a strong contender to be used. Despite this there was in fact only one mention of geophysics in PPG16 ('Developers may wish to carry out geophysical surveys as part of their own initial archaeological assessment'). While this was hardly a ringing endorsement, that mention paved the way for regular consideration in archaeological evaluations.

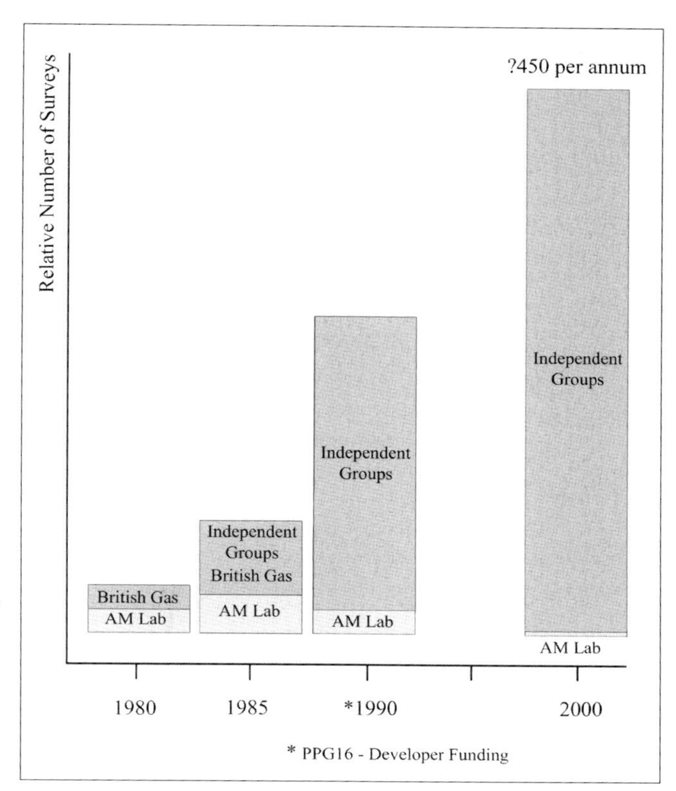

3 *A graph showing the relative increase in geophysical surveys for archaeology within the past two decades. The dominance of the independent, or commercial, groups in recent years is due to a change in the source of finance from public to private funds*

As a result of this increased demand, more commercial groups were required to offer a geophysical service. Today it is likely that somewhere in excess of 450 surveys are undertaken each year in England alone. Nearly all are carried out by commercial firms and it is worth noting that the percentage undertaken by AM Lab has reduced from *c.*40 per cent in 1985 to less than five per cent today. Yet, as stated at the start of the chapter, the number of people involved in archaeological geophysics has never been greater. It is likely that about 20 groups are actively undertaking commercial work in archaeological geophysics in Britain, though over 75 per cent comprise one or two employees.

In Europe new centres of research in archaeological geophysics are flourishing in Austria, France, Germany, Holland, Italy and Switzerland. Greece and Poland continue to have active groups and work is widespread through many eastern European countries. In Japan the use of ground-penetrating radar tends to predominate, as is the case in the States, where other geophysical techniques are beginning to be employed to an ever-increasing extent. There are still very few published articles on archaeological geophysics in South America, Africa and Asia as a whole, but this is more a reflection on the nature of archaeological research in these continents. At present the richer countries are still at the forefront of research; however, the geographical spread of countries where the techniques have been applied is ever increasing. A cursory glance at recent articles published in *Archaeological Prospection* reveals case studies from Albania to Zimbabwe, Australia to the USA and China to Wales!

The integration of science into archaeology is now routine and nearly every university archaeological department in Britain now teaches archaeological science, including geophysics. The dedicated student can also undertake an MSc in Archaeological Prospection at Bradford, or a course in Aerial Photography with Geophysical Survey in Archaeology at Glasgow. The fact that so many archaeologists are taught aspects of geophysics is an important factor in its widespread usage. In effect, archaeologists feel comfortable with the techniques and have little problem with specifying that such work should be carried out. This is especially important when the archaeologist works within a planning department of a local authority and has the responsibility for checking the quality of field reports.

The last decade of the twentieth century saw the publication of two textbooks on archaeological geophysics: Tony Clark's *Seeing Beneath the Soil* (1990, revised 1996) and Irwin Scollar's *Archaeological Prospecting and Remote Sensing* (1990). The former is popular, the latter more theoretical, in their respective approaches to the subject. In fact, the books very much reflect the twin avenues along which geophysics was developing at that time and to some extent continues today. In 1991 the Institute of Field Archaeologists published a technical paper on the role of geophysics in archaeological evaluations (Gaffney *et al.* 1991, revised in 2002) and in 1995 English Heritage issued 'guidelines' for archaeological geophysics (David 1995). Since 1994 an international journal, *Archaeological Prospection*, has been published by Wiley, and the opportunity has arisen again for archaeological geophysicists to have articles published in a subject specific peer-reviewed journal. The journal is targeted at anyone involved in archaeological geophysics, from theorists to practitioners. However, it is a rather sad reflection on the majority of commercial archaeological geophysicists that they have not seized the opportunity to publish their material. It is likely that geophysical reports will be submitted for inclusion in a county Sites and Monuments Record and they will receive no wider review. It is perhaps ironic that as the amount of geophysics conducted for archaeological purposes increases, a smaller range of the work is seen by the general public.

The Megger Earth Tester used by Atkinson is an instrument that has long been assigned to the museum shelf. Made from stout materials, many no doubt lasted longer than those who surveyed with them. The hand-cranked device served its purpose and made way for the transistorised Martin-Clark, and the constant current AC devices such as the Bradphys and the Geoscan Research RM instruments. Automated systems, some attached to tractors or quad bikes, are now in existence. A seemingly relentless progress has produced equipment that is robust, lightweight, and ready for survey. Having reached the twenty-first century, archaeological geophysics is faced with many exciting challenges, some technical, some methodological and some political.

2

TECHNIQUES OF GEOPHYSICAL INVESTIGATION: SOME PRINCIPLES

Introduction

In this chapter we will consider the basic principles and theories for a variety of geophysical techniques used in archaeology; in effect we will try to reveal the science behind the black box. The aim is to provide the reader with an introduction to how the techniques work, and we make no apologies for trying to disentangle often complex subjects in a simplified form. Those wishing to pursue more detailed avenues will need to consult the numerous references provided. The standard geophysical textbooks of Kearey and Brooks (1991), Musset and Khan (2000), and Telford *et al.* (1990) all give excellent descriptions and explanations of the theories involved.

The term geophysical survey is employed here to indicate those techniques that use physical measurements at the surface of the Earth to investigate shallow depths, normally down to a few metres. Nearly all of the techniques used for archaeological purposes have been borrowed from geological geophysics, although in some cases the implementation has been radically changed. While many techniques have been tried, only a handful have been successful in detecting archaeology on a more or less regular basis: a list of the most important can be seen in figure **4**.

The types of instruments that are used can be classified in a number of ways. For example, some need to be inserted into the ground, others simply require contact with the ground while some of the instruments are carried above the surface. However, the most important difference between the techniques is whether or not they induce a phenomenon or if they measure what is there directly: they are termed 'active' and 'passive' respectively (**4**).

Techniques of Detection in Archaeological Geophysics

Method	Active or Passive	Frequency of Use
Electrical Resistance	Active	High
Magnetometry	Passive	High
Electromagnetic	Active	Mid / Low
Magnetic Susceptibility	Active	Mid / Low
Metal Detectors	Active	Low
Ground Penetrating Radar	Active	High / Mid
Seismic	Active	Low
Microgravity	Passive	Low
Induced Polarisation	Active	Low
Self Potential	Passive	Low
Thermal	Passive	Low

4 *A list of geophysical techniques used for locating, delimiting and investigating archaeological sites*

Resistance or resistivity survey

The basis for this method is that electric currents are fed into the ground and the resistance to the flow of these currents is measured. Where they 'meet' buried wall foundations high resistance readings are recorded, while if silted-up ditches (which tend to be wetter than the surroundings) are encountered low resistance readings ensue. By mapping zones of high and low resistance it is possible to identify, for example, the layout of buildings or the size and orientation of a ditched enclosure.

What can you detect with this technique?

Interpreting data from resistance surveys can be very difficult: it is the multiplicity of factors that affect the moisture near the surface that can make this data quite daunting. However, the basic concepts regarding high and low resistance features are fairly straightforward and are as follows:

High resistance anomalies	**Low resistance anomalies**
Walls	Ditches/pits
Rubble/hardcore	Slots and gullies
Made-up surfaces	Drains
Roads/trackways	Graves
Stone coffins/cists	Metal pipes

The problem with resistance data is that the 'normal' response can vary with the season. For example, during winter there may not be any contrast between a ditch and the surrounding saturated soil. The 'signal' from the ditch will blend into the background variation. Another difficulty may arise if the width of a ditch is excessively large. Under these circumstances the ditch can act as a huge sink with the moisture gravitating to the bottom, leaving the top relatively dry, especially in soils with high clay content. If the ditch is large by comparison to the separation of the probes used in the resistance array, then the response can actually reverse and the ditch can give a high resistance response. This has been shown to be the case in surveys over the vallum ditch on Hadrian's Wall (**5**). The graph shows a fairly uniform response throughout the length of the traverse apart from the clear high resistance anomaly. Excavations showed the ditch to be flat-bottomed, with steep sides *c*.7.5m wide and 1.6m deep and the fill was much drier than the surrounding water-logged ground.

It is rare for a normally high resistance body such as a building to show as a low resistance anomaly, except when water is 'puddling' on top of an impermeable surface. However, there are plenty of examples of buried walls being invisible to this technique. Sometimes this may be the result of the size of the feature with respect to the volume of the earth that is sampled, which is itself a function of the probe separation, but more likely it will be due to a lack of moisture contrast between the feature and the dry soil that surrounds it. The variation in moisture content can change very rapidly even over a small area and this makes the interpretation of resistance data highly challenging.

Principles

The method draws on the ability of the soil to allow electric current to pass through it. This property is directly related to the interstitial water that is held in the soil and the various salts that may be present. The current flowing through the material is in proportion to the potential difference, or voltage, that is used. This

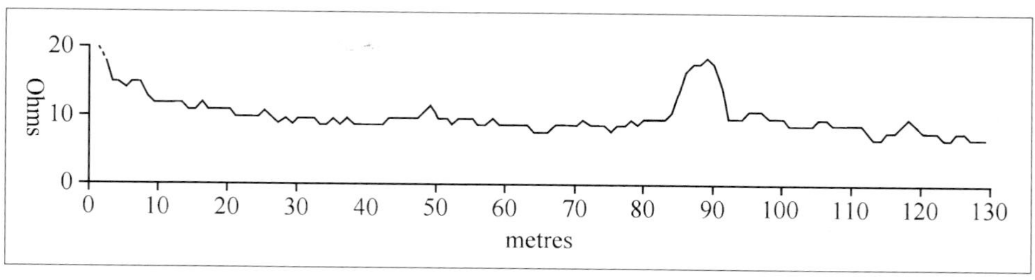

5 *The results of resistance traverse over the vallum ditch on Hadrian's Wall. Although ditches normally result in lower resistance values than the local average, large ditches, such as the vallum often produce reverse responses. It is believed that in this case the top of the ditch has dried preferentially as a result of water draining to the lower part of the feature. The survey was conducted with a Bradphys resistance meter using the Twin-Probe configuration. Readings hand logged every 1m along a series of widely spaced parallel traverses.* After Gater, 1981

relationship, called Ohm's Law, applies to all the methods in this section. The resistance (R) can be established by measuring the current (I) (measured in amperes, or amps) flowing through a body of material and monitoring the change in voltage (V) (measured in volts) across the material:

$$R = V/I$$

If the current is kept constant then the resistance can be worked out by monitoring the change in voltage. The resistance is expressed as ohms which is often shown with the symbol Ω.

It is important to understand that resistance is a 'bulk' measurement that relates to the whole of the material through which the current is flowing – that is both the intrinsic properties of the substance and how much of it is present. Generally a more informative property is resistivity (ρ). This can be considered as a measure that is due to the material itself and will not change with the amount or shape of that material. Resistivity is measured in ohm-m and is the inverse (or opposite) of conductivity, just as resistance is the opposite of conductance.

Meters used in practical situations measure resistance rather than resistivity since resistance is probe-geometry dependent. For most applications involving gridded area survey, plotting variations in 'earth' resistance is quite satisfactory – in others where modelling or comparisons are made between different probe arrays, resistivity should be computed. To avoid confusion a report of an electrical survey should be clearly marked as either resistance or resistivity.

In order to measure the resistance of the earth a current is made to pass between two electrodes inserted into the ground. These are termed the 'current' electrodes and an alternating current supply is 'injected' through these points. Alternating current (AC) is preferred to a direct current (DC) supply as this avoids spurious polarisation voltages being set up around the probes by electrolytic action. Modern resistance meters maintain a constant value for the current. This means that the instrument can avoid the need for an ammeter to measure the current and can be calibrated to read resistance directly as it is proportional to the voltage.

In practice four electrodes are used; alongside the two current probes, two further probes sample the voltage (resistance) between these two points. This is carried out because the current probes have a finite (and unknown) 'contact' resistance with the earth. Although 'contact' resistances also occur at the second set of probes, the values are usually considerably smaller than the input impedance within the voltmeter itself and therefore do not affect its reading.

It is worth noting that the four electrodes can be configured in many different ways, although only a small number of variations are commonly used in archaeology (**6**). The configurations are normally termed 'arrays' and are either named after notable pioneers of geophysics, such as Schlumberger or Wenner, or are descriptions of the geometry, such as 'Double Dipole'. Depending on the array the response to the same buried feature can vary – fortunately this is in a predictable manner.

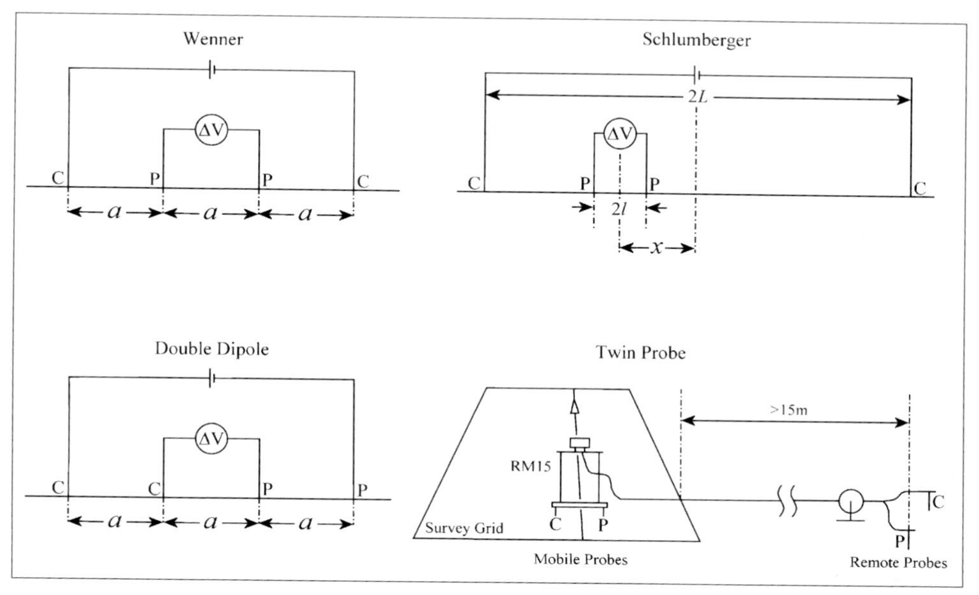

6 *Four resistance arrays used for archaeological prospecting. For the Wenner the inter-probe separation, a, is equidistant between each probe and all four probes are effectively moved together. In the version of the Schlumberger shown here, the distance 2L is much greater than 2l and only the potential probes are moved for each reading. The most common array for archaeological purposes is the Twin-Probe and the distance between the mobile and remote pair of electrodes is at least 30 times the separation between the probes on the mobile frame, which in the example shown is 0.5m*

Probably the easiest array to understand is the 'Wenner' where the two potential probes are situated within the current probes and they are equidistant from each other (**7**). This is the arrangement used by Atkinson back in 1946. The Wenner array is simple (but slow) to use for archaeological survey where readings are often taken every metre across a grid. However, the form of response, that is the shape of a traverse of readings, can be highly complicated as all four probes move across the target – this often leads to multiple peaks over a single feature.

During the 1960s/70s a great deal of research was carried out into ways of simplifying resistance survey. For example, those using the Wenner array employed a rigid frame on which all probes were attached and could be moved by one person. A similar rigid frame approach was developed by proponents of the 'square' array, which had probes on each corner and had the added benefit of the resistance meter conveniently mounted on a platform between the probes. This board also allowed space for the readings to be written down. A third configuration, developed largely in Britain and now the most favoured probe array is the so-called 'Twin-Probe' (**6, 8 & colour plate 1**). This array can be regarded as a Wenner configuration that has been split in two, thus there are two current-potential pairs, one 'mobile' and the other 'fixed' at a distance. Although this array was first experimented with by Schwarz (1961) working in Switzerland, it is commonly accepted that the Bradford

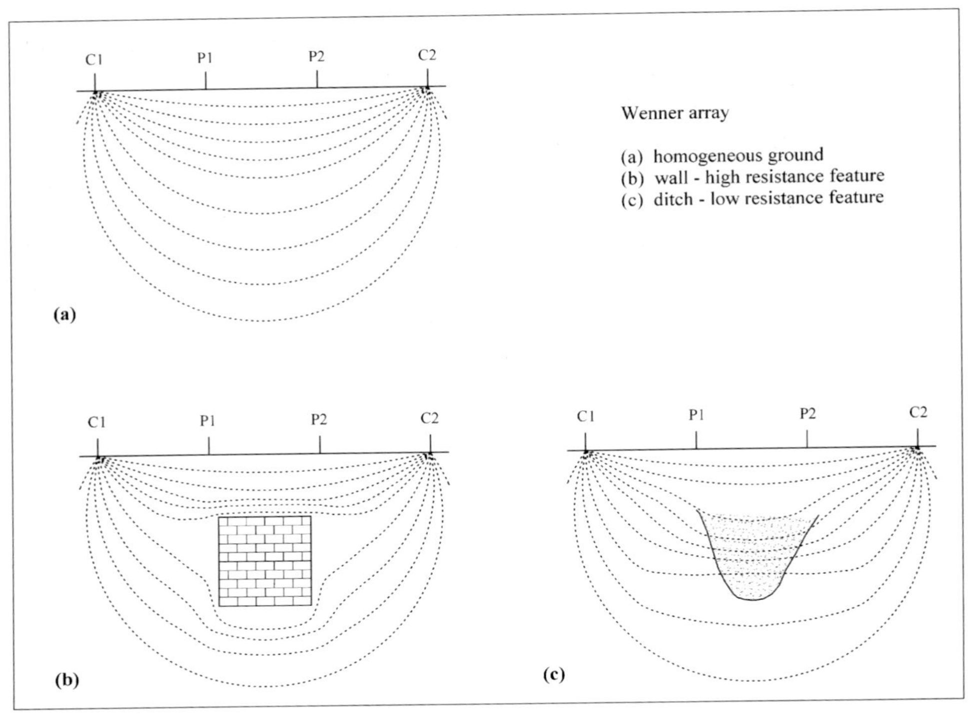
C1
P1
P2
C2
Wenner array
(a) homogeneous ground
(b) wall - high resistance feature
(c) ditch - low resistance feature
(a)
(b)
(c)

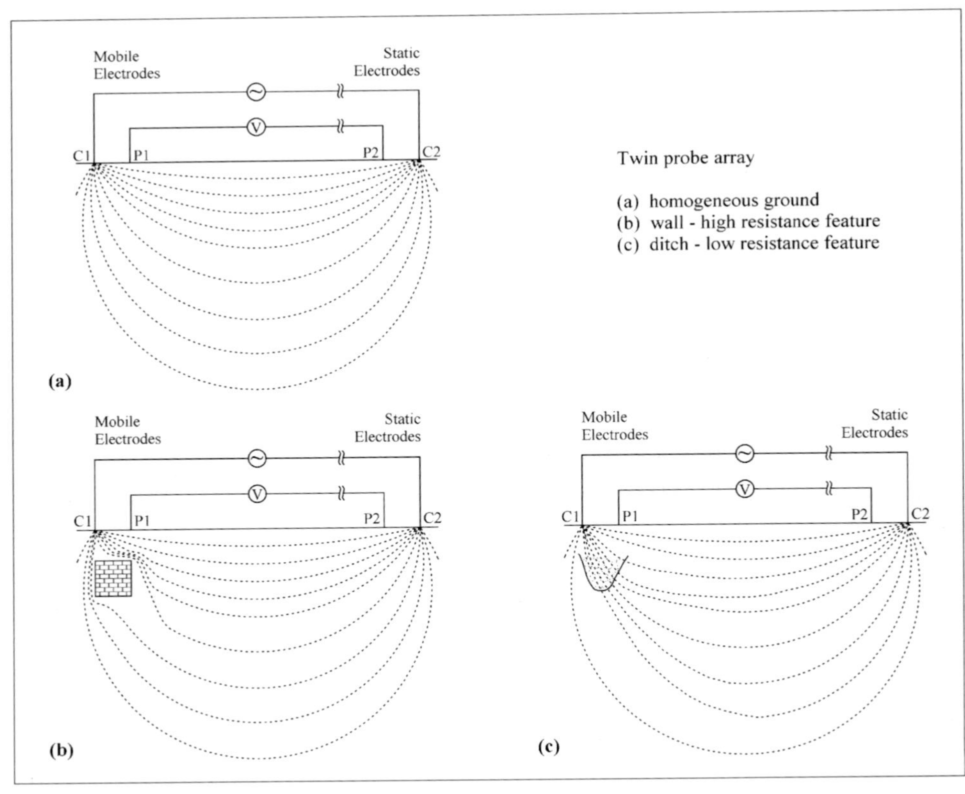
Mobile Electrodes
Static Electrodes
C1
P1
P2
C2
Twin probe array
(a) homogeneous ground
(b) wall - high resistance feature
(c) ditch - low resistance feature
(a)
(b)
(c)

University-based archaeological scientist Arnold Aspinall was the most important advocate of the array, and a seminal publication first documented the potential in archaeological terms (Aspinall and Lynam 1970). It was clear at that point that the array had a number of disadvantages as well as advantages over the Wenner.

Manipulation of the basic resistance equation shows that the value for the resistance at a particular location should be half if measured using a Wenner as opposed to the Twin-Probe (**9**). However, the percentage change on traversing a buried target can be as great as 50 per cent with the Wenner but as little as 15 per cent with the Twin-Probe. So what then are the properties of the latter that make it the most popular array used in resistance surveys for archaeology?

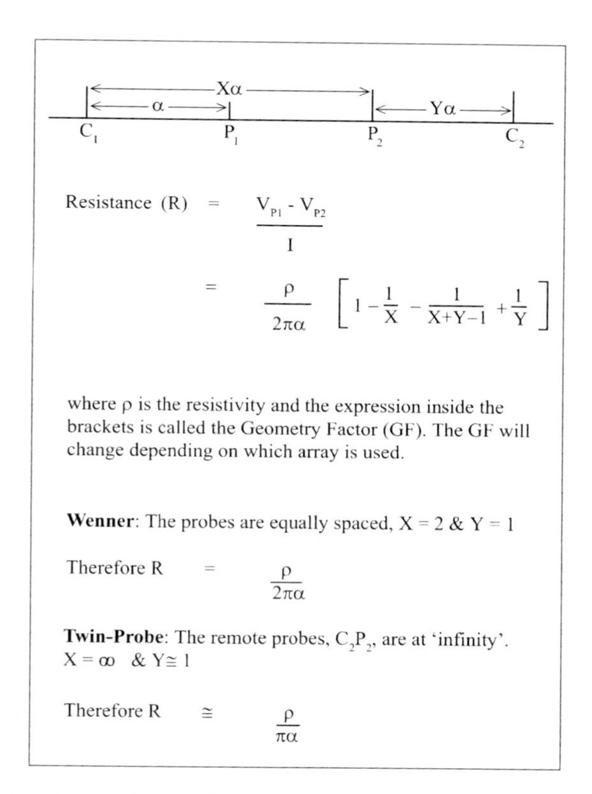

7 (Opposite above) *Schematic diagram of a measurement taken using the Wenner array showing how the electrical current flows a) through soil devoid of features b) around a highly resistive 'wall' c) through a soil feature of lower resistivity than the surrounding material*

8 (Opposite below) *Schematic diagram of a measurement taken using the Twin-Probe array showing how the electrical current flows a) through soil devoid of features b) around a highly resistive 'wall' c) through a soil feature of lower resistivity than the surrounding material*

9 (Above) *The basic equation for resistance survey. Knowledge of this equation allows a prediction to be made regarding the relative strength of the response for each array or allows the results to be converted from resistance to resistivity for comparative purposes*

The array is characterised by two separate pairs of probes (C1P1 and C2P2 or AM:BN) in which the distance between the mobile probes and the static or 'remote' probes is said to be 'large'. In practice this needs to be at least 30 times the distance between the probe separation on the moving frame – this multiple is said to place the remote pair effectively at infinity. This means that for a survey using a 0.5m separation the remote probes should be at least 15m away from the nearest point on the survey. If the distance is less, then the value measured by the mobile probes will be influenced by their relative position to the fixed probes. Assuming that the '30 times' rule is maintained then it can be shown that the distance between the probes will only create a variation of about 3 per cent in the measured value of the resistance. This is important as it is less than the typical value of archaeological anomalies (10–20 per cent). A further implication is that the mobile probes do not need to be maintained in the same orientation with respect to the remote pair. This feature allows zigzag data collection as well as some movement to allow for a physical obstruction at the position of measurement (**10**). It should be remembered that each measurement is effectively a sample of the earth at that location and that the surveyor should feel confident about getting the measurement position within *c.*10cm.

The rule of thumb in resistance survey is that the wider the separation, the deeper the array will see into the ground (see the later section dealing with VES and pseudosections). In the case of the Twin-Probe it is not the overall spread that matters but the distance between the current and potential probes in a pair. Of importance is the fact that the Twin-Probe taps into a volume of ground where the potential gradient is changing rapidly next to the current probes and this ensures that the majority of the variation in signal results from very near surface inhomogeneities. A 0.5m Twin-Probe is likely to respond to features at a maximum depth of about 0.75m. In many parts of the world this depth encompasses much of the archaeology in rural locations. The array is also less liable to reflect geological variation than many of the other resistance arrays.

The final advantage that the Twin-Probe has over the Wenner is that the response is normally a single peak rather than a complicated multi-peak. This means that the Twin-Probe data will have better resolution than the Wenner. This is clearly important on complex sites where walls and ditches may be re-built or re-cut in close proximity to each other.

Theoretical aspects of the probe arrangement have already been discussed, but what happens when the survey extends beyond the initial grids and the cable reaches its maximum extent? When this occurs the remote probes have to be repositioned, again using the '30 times' rule with respect to the grid squares to be surveyed. This is achieved by taking a reading and moving the remote probes to a new location, but keeping the mobile probes stationary. At the new remote location the probes should be inserted into the ground with a similar separation as before, that is approximately 0.5m. The reading on the meter will be different to the previous one as the resistance of the ground beneath the remote probes will have changed. The distance between the remote pair should be varied until the

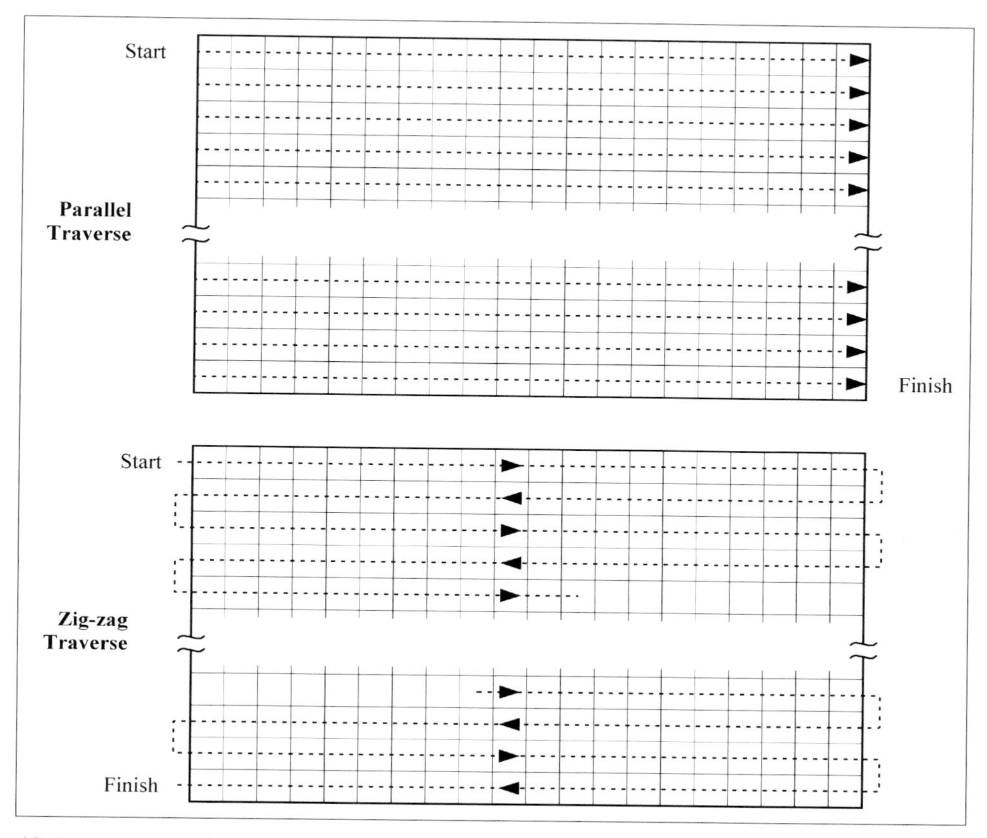

10 *In area surveys, data can be collected either in parallel or zigzag formation. Parallel is much slower but can reduce the number of errors in the data set. Whichever strategy is used, this information will be required by the computer software to place the readings in their correct position within the grid*

'new' reading is the same (or very similar) to the one noted at the old remote position; this is known as 'normalisation' (**11**). The new grid can now be surveyed. It is important that this procedure should be carried out before data collection starts on a grid rather than part way through. A number of theoretical issues should be noted here. To be a true Twin-Probe the distance between the C1 and P1 should be the same as C2 and P2. If the distance between the remote pair (C2 and P2) has varied greatly as a result of the normalisation process then the Twin-Probe assumptions will have been violated. In practice this matters very little, although this variation does make the conversion to resistivity problematic, if not impossible. It is good practice to ensure that the C2P2 separation lies between 0.5x to 4x C1P1 separation. Some surveyors prefer to extend the distance between the remote probes to as great a distance as possible (e.g. Dabas *et al.* 2000). The main reason for this alternative method is that the remote probe contribution to the overall resistance will decrease and the observed reading will be a more accurate estimate of the resistance of the volume of material beneath the mobile probes.

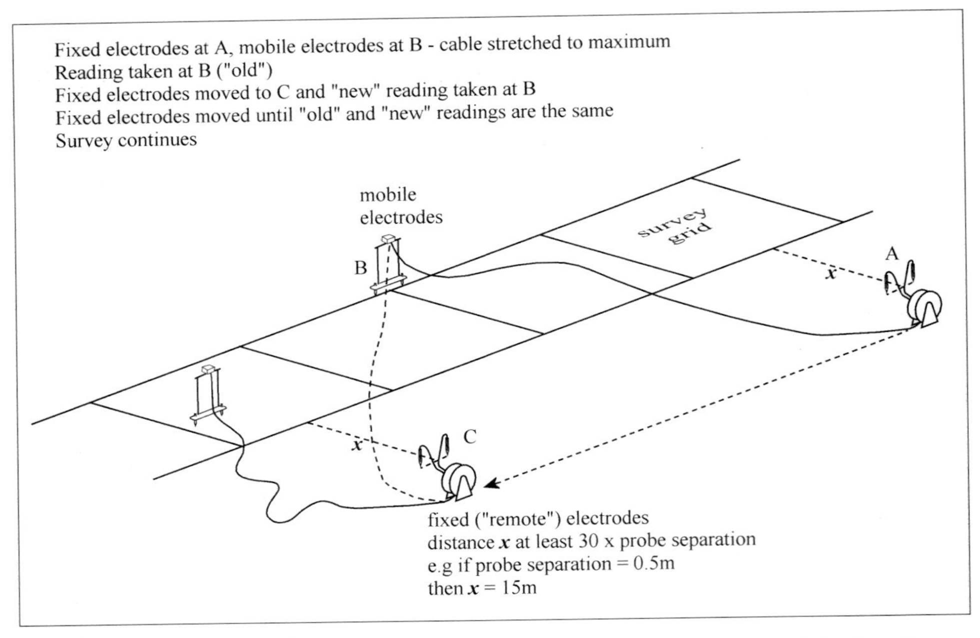

11 *When undertaking an area survey with the Twin-Probe it is necessary to survey the whole of a grid square without moving the mobile probes. If that cannot be done then the probes should be moved prior to the survey of the grid. This must be achieved as outlined in the diagram and is called 'normalisation'. The most important fact is that the 30 times rule must be obeyed i.e. for 0.5m separation between the probes on the mobile frame the minimum distance the remote probes can be from the grid square (x on the diagram) is 15m.* After Musset and Khan, 2000

Taken to extremes this variation becomes a different array know as the Pole-Pole, or Bipole. However, care must be taken when using the wide separation to ensure that the remote probes are sufficiently far away from the mobile probes. One practical benefit is that normalisation is not usually required when the remotes are moved, as the influence of the 'remote' probes on the resistance measurement substantially reduces. Of course the increased length of cable can be a considerable inconvenience in surveying with a Pole-Pole system.

Vertical electrical sounding (VES), pseudosections, electrical imaging and tomography

The majority of resistance surveys undertaken for archaeological purposes are performed with a constant separation between the probes and the data are collected at sample intervals over a regular survey grid. Normally the separation is sufficient to measure through the topsoil and to sample the archaeology buried below, usually within 1m of the ground surface. The distance between samples is a balance between the expected size of the likely features and the time that the surveyor has to investigate the area; for most applications the distance is 1m, and while smaller sample intervals are not uncommon it is unusual to increase the distance beyond 1m. While this area survey approach is routinely the most efficient way to investigate a site, the problem is that many assumptions have to be made, particularly about the depth of the potential archaeology. An alternative approach therefore is to look successively deeper into the ground.

Geologists have investigated horizontal strata of differing resistivities by a technique called Vertical Electrical Sounding (VES). In this technique the separation of the array is expanded around a central point and the resistance is measured for an increasing depth of ground. A result of this expansion will be a change in resistance; in fact the change is a product of both the separation and the changes in the underlying strata. A mathematical correction can be applied to compensate for the variation in separation and, after analysis against standard curves, an assessment can be made of the layering of the ground at that location. In practice this technique has rarely been used in archaeological situations. However, the principle of expanding the probe separation is very important in the measurement of pseudosections (also called electrical imaging), which involves VES-type measurements at many points along a traverse.

In generating a pseudosection a long line of electrodes are set out, typically 25 or 32 in number, at regularly spaced intervals, usually 1m for shallow investigations and up to 5m for deeper work (**12**). A box of electronics is connected to the electrodes by a multi-core cable and a built-in computer usually drives the switching and data recording processes. A series of readings, typically either Wenner or Double Dipole, are taken using different probes along the line. A vertical profile of measurements is generated by expanding the probe separation at all possible array centres. The resistance for each measurement point is calculated and corrected to resistivity. The value is placed at a nominal depth, which in the case of the Wenner array is equal to the separation between the probes. However, it is clear that gauging the depth of the measurement point is not an exact science. As a result of this arbitrary depth estimation, the generated image of the resistivity is known as a pseudosection. A rough pseudosection can also be collected using the standard Twin-Probe array with an additional multiplexer (see chapter 3).

At first glance a pseudosection can be 'read' like a drawn section from an archaeological excavation. However, at this stage the data are 'approximate' in that there are a number of distortions resulting from the changing relationship between the probes and the buried archaeology and strata (Aspinall and Crummet 1997). To establish an image that is closer to the real variation in resistivity, a computer program is used to generate a model data set based on the measured pseudosection (Barker 1992). This model effectively produces a 'true' depth and can be corrected for topographic variation. This process is usually called Electrical Imaging, although it is a form of tomography similar to that developed for medical purposes that utilise X-rays (Szymanski and Tsourlos 1993).

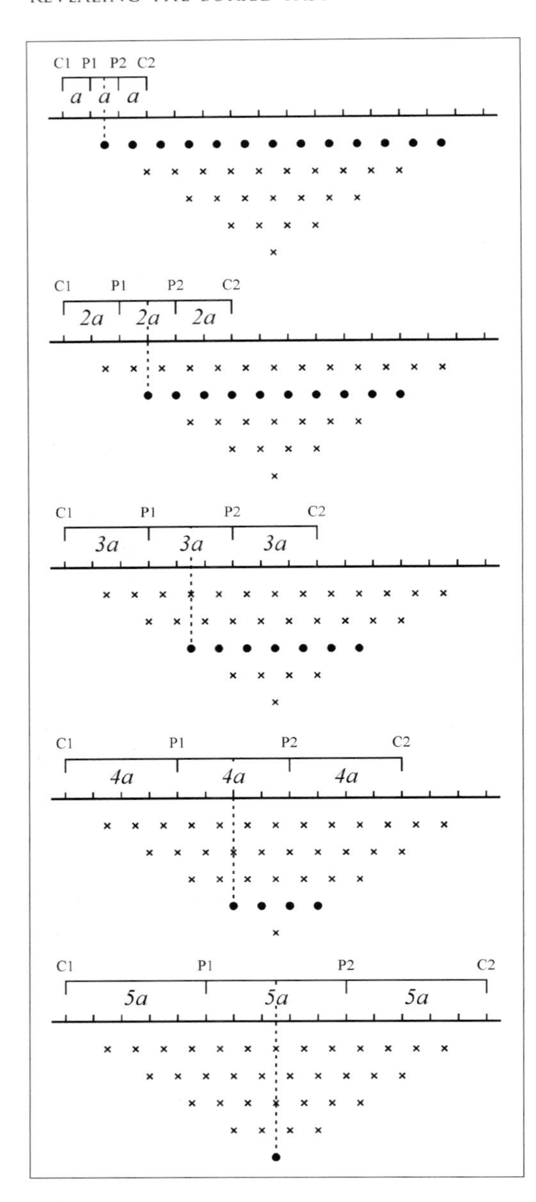

12 *An electrical pseudo-section can be generated by undertaking a number of surveys with differing probe separations along the same traverse. In practical situations the measurements are taken using a large number of static probes that are automatically switched 'on' or 'off' by computer-controlled software. The greater the separation between the four measuring probes the deeper the investigation. The black dots represent readings at each level as the array traverses from left to right*

Magnetic survey using a magnetometer

Magnetic survey methods rely on the ability of a variety of instruments to measure very small magnetic fields associated with buried archaeological remains. In effect a ditch, pit or kiln can act like a small magnet, or series of magnets, that produce distortions (anomalies) in the Earth's magnetic field. In mapping these slight variations, detailed plans of sites can be obtained with the advantage that the buried features often produce reasonably characteristic anomaly shapes and strengths.

What can magnetometers detect?

When Aitken first used a proton magnetometer to locate archaeological features at Water Newton during 1958, he was targeting heavily-fired material in the form of kilns of Romano-British date (Aitken 1986). The magnetometer detected the magnetic changes that result when firing reaches high enough temperatures to alter the magnetic properties of the kiln's clay matrix structure. However, the magnetometer also detected much weaker enhancements from pits and ditches which were the product of enhanced magnetic susceptibility.

It has subsequently been demonstrated that magnetometers are particularly good at detecting fired or burnt features, such as kilns, ovens and hearths. More routinely they respond well to negative features, i.e. ditches and pits, cut into the subsoil. They are ideal for locating industrial areas although it is sometimes difficult to differentiate between those of 'modern' and 'ancient' origins. In the hands of an experienced operator these instruments can be used to scan out specific anomalies or areas of interest. They are not very good at detecting structures (buildings) unless they are made of fired brick or have been burnt or buried in more magnetic soil.

Principles

The basis for magnetic prospecting is the presence of weakly magnetised iron oxides in the soil. Depending on the state of iron oxides, the material will exhibit either a weak or a strong magnetisation. There are two phenomena that are relevant to the discussion of magnetic anomalies:

1. Thermoremanence describes weakly magnetic materials that have been heated and have acquired a permanent magnetisation associated with the direction of the magnetic field within which they were allowed to cool. For this to be possible the material must first of all be heated so that its temperature passes through a specific value, known as the Curie Point. The Curie Point varies depending upon the minerals that are present. For haematite it is 675°C while it is 565°C for magnetite. In exceeding these temperatures the iron content of the clay is demagnetised and effectively wipes clean the magnetic properties of the material. As it cools it is re-magnetised and acquires a new and permanent magnetic property specific to its relative position in the Earth's magnetic field. In fact the magnetic minerals become aligned with the Earth's magnetic field at the time of cooling and this is called thermoremnant magnetism. This fundamental change in the properties of the minerals is often linked with substantial chemical and physical modifications. Archaeological features that have been through this mechanism include baked clay hearths, and kilns used for ceramic manufacture. Importantly, whilst this creates a readily identifiable, even characteristic, signal, this is the same property which makes magnetic results on some igneous geologies so difficult to interpret.

2. The magnetic susceptibility of a material is defined in terms of the magnetism induced in a sample when it is placed in a magnetic field. The more magnetised a material becomes the higher the susceptibility is said to be. It is important to know that magnetic susceptibility is a temporary response that can only be measured when a magnetizing field is present. However, as the Earth's magnetic field is always 'on' then it follows that magnetic susceptibility is also always 'on'. Consequently a 'passive' detection device (such as a magnetometer) can be used to measure the changes in susceptibility as well as an 'active' instrument that induces a magnetic field (such as a field coil).

Magnetic susceptibility is the key in the production of coherent results from magnetic surveys. Moreover, not only can the difference in magnetic susceptibility between topsoil and subsoils be used in a predictive manner for the success or otherwise of magnetometer surveys, but the spatial variation of the property throughout the topsoil can itself be an important indicator of 'activity' in the archaeological context.

The major contribution in describing the possible mechanisms at work in the enhancement of magnetic susceptibility is that of Le Borgne (1955, 1960). He suggested a simplified transition, via conditions of reduction followed by oxidisation, as follows:

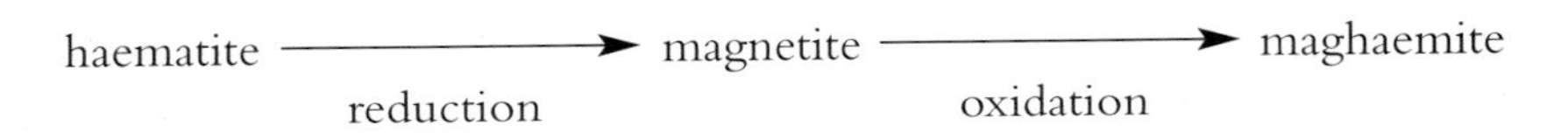

In fact Le Borgne proposed two mechanisms which produce these conversions: burning and the fermentation effect. The burning mechanism is fairly well understood and hinges upon the thermal alteration of weakly magnetic/antiferromagnetic iron oxides to more magnetic oxide forms. By way of contrast the fermentation pathway is the source of considerable debate. The so-called fermentation effect is important as it helps explain one of the key factors in magnetic survey; that is this effect offers a route for topsoils to have a higher magnetic susceptibility than subsoils, assuming a non-igneous parent. Le Borgne's mechanism requires alternating reduction and oxidation states and is probably an oversimplification of the process. In fact it is now believed that the mechanism is highly complex and involves the interplay between biological-pedological systems and probably the interaction of microbia, soil organic matter and soil iron (Maher and Taylor 1988; Fassbinder *et al.* 1990). The important fact is that even in the absence of the heating mechanism of enhancement, detectable features can be produced, for example, by the infilling of a ditch with relatively enhanced topsoil materials.

Furthermore, general anthopogenic activity, in the form of settlement and rubbish disposal practices, increases susceptibility and has an immediate influence on the detection of buried features (Tite and Mullins 1971). The exact pathway for the increase in magnetic susceptibility in these circumstances is probably a fusion of several mechanisms. Indeed, low temperature 'burning' in the form of experimental short-term camp fires have been shown to produce significant increases in magnetic susceptibility (Linford and Canti 2001). What is certain is that a magnetic contrast must exist between a feature and the surrounding soil and for a magnetometer to detect an anomaly produced by the feature: this contrast can be the result of thermoremnant magnetism, magnetic susceptibility or both. Whichever of the phenomena supplies this contrast from the background, they both depend on the original concentration of iron oxides in the soil. Therefore, soils with low iron oxide content will only produce small magnetic anomalies as the maximum amount of magnetisation is constrained by the level of iron oxides. On such soils the enhancement in magnetic susceptibility will be reached that cannot be passed no matter how much anthropogenic activity is undertaken on a soil.

Fortunately, the instruments that are available for archaeological survey, known as magnetometers, are extremely sensitive, and even features producing very weak anomalies can be detected. Changes as small as 0.1 nanoTesla (nT) in an overall Earth field strength of *c.*48 000 nT can be readily detected using a dedicated instrument. In addition, the mapping of an anomaly in a systematic manner will allow an estimate of the type of material present beneath the ground, as well as some estimate of burial depth.

In terms of magnetic detection of archaeological features, it is easiest to imagine the situation of a ditch cut into subsoil. Anomalies that are of interest are often simply the product of the relative contrasts between the subsoil and the magnetically enhanced topsoil that fills the feature. The silting or deliberate infilling of the ditch results in an increase in iron-rich material from the topsoil. This can create a characteristic anomaly which changes in shape depending on the interaction of the localised field with the Earth's magnetic field. In Britain, a pit or ditch containing enhanced deposits will produce an anomaly relative to the Earth's field with a positive (high) peak to the south and a corresponding negative (low) to the north of the feature. Depending on the location of the survey on the Earth, the lateral displacement between the positive peak and the centre of the feature is usually no more than 0.25m, which is often close to the precision on the sampling interval.

It is worth stressing that whatever magnetic properties can be attributed to a particular feature, these properties are embedded within the complex magnetic field that is created by the Earth. The origin of this field is very complicated and gives rise to major, or large-scale, deviations from the simple bar magnet picture that is often portrayed. In fact there are many localised distortions due to high iron-bearing rock, such as is found in igneous regions. There are also time-dependent variations in the Earth's field. One of these variations, called 'secular', is sufficiently long term that it forms the basis for magnetic dating. Smooth variations on a daily basis that result from the movement of the Earth with respect to the Sun and the Moon are called 'diurnal', and even shorter term variations can be caused by magnetic storms, where a sharp decrease in the field will be followed by a slow rise. Some instruments will be affected by such variations, thus changes have to be removed before a 'true' anomaly signal can be obtained. This can be a tedious business and requires significant monitoring and processing.

Fluxgate magnetometers

In Britain, and in many other parts of the world, the most common type of magnetometer that is used for archaeological investigation is based on the fluxgate sensor. These sensors are made from a highly susceptible metal core, traditionally 'Mumetal', that saturates in the Earth's magnetic field. Around the core are wound two coils, known as the primary and the secondary. An alternating current is driven through the primary and the core is driven in and out of saturation. In fact the saturation flips from one polarity to the other with the change in current. This change induces a current in the secondary coil, the frequency of which is linked to the rate at which saturation is induced in the core. If a magnetic field, in this case the Earth's, is applied to a sensor then the rate of core saturation will alter as there will be preferential magnetisation of one polarity, keeping the core saturated for longer in one direction. As the core pulses between the saturation polarities the external field can induce an electrical pulse in the secondary coil. This pulse is directly related to the field strength of the external field.

There are a number of points relating to this type of instrument. Probably the most important aspect of the fluxgate is the fact that it is directionally very sensitive along the axis of the sensor core. The instruments are usually held with the axis vertical resulting in the measurement of the vertical component of the magnetic field. The response of these instruments to the same feature will vary significantly over the surface of the Earth. This is because the vertical field is small near the equator and large near the magnetic poles. For example, when surveying near the magnetic North Pole the response is substantially a positive anomaly over the centre of the feature. This positive anomaly decreases in strength and moves slightly south of the feature if its location is nearer the equator. At the equator the response is an equal positive/negative centred above the feature.

Fluxgate instruments are usually said to be configured in what is known as gradiometer mode. For such a system the two sensors are maintained vertically above one another, normally 0.5 or 1m apart. The top sensor measures the Earth's magnetic field at that point while the lower sensor measures the same field but is affected by any buried feature that is closer to it. By removing the first from the second then the anomaly due to the buried feature can be easily calculated. If there is no localised variation due to buried archaeology, or modern surface rubbish, then the value measured on both sensors will be the same and the difference between the two will be zero, or at least as close to zero as instrument and soil noise allows. It is much easier to monitor the variation in the background in this fashion than to try to measure absolute changes, which are very small by comparison to the Earth's magnetic field. As the measurements and calculations can be concluded in a fraction of a second, the instrument can give real time values and it can be regarded as a continuous reading instrument. A plus with this type of instrument is that no fancy processing is required to correct for the various short- or long-term changes to the Earth's magnetic field as both sensors are equally affected by these factors. Another positive aspect of working with a gradiometer is that a uniform background will give a value of zero as the instrument has already filtered out broad and slow varying change due to geological or similar variation. This inherent filter therefore reduces the need to process data prior to producing an acceptable archaeological image. Evidently, the narrower the sensor separation, the shallower the search depth of the instrument. Therefore if survey areas are believed to contain alluvium or other types of built-up soil, then a fluxgate system with the greatest available separation should be considered. However, for a fluxgate gradiometer to work efficiently it has to be mounted within a rigid system, with the manufacturer dictating the sensor separation. If fluxgate data have to be collected at differing depths then simply changing the distance between sensors on an instrument is not practical: normally a different instrument must be used or an alternative set of sensors must be swapped into a modular system.

A true gradiometer requires the sensors to be very close together, probably closer than can be usefully maintained for archaeological purposes. In most commercial instruments the values are measured in nT ('gamma' in older units) rather than nT/m as would be expected from a true gradient. In doing so the manufacturers acknowledge that they are really measuring the difference between the fluxgates rather than the gradient of the vertical magnetic component. While this may seem a trivial point, it is worthwhile noting the measurement unit and sensor separation before analysing the response between different instruments.

Caesium magnetometers

In the early 1960s and in parallel with the development of fluxgate systems, a class of instruments known as caesium or alkali vapour magnetometers came into vogue. A pioneer of this instrument was Elizabeth Ralph who collected data on nearly 50 sites throughout the world. She will be best remembered for her painstaking work at the Greek colony of Sybaris, where she worked for about two years and collected an amazing *c*.400,000 readings that were written down and manually graphed. She did not work with the instruments in the gradiometer mode, but used single sensors that are influenced by geology and also deeply buried archaeology. She was successful at locating concentrations of roof tiles made in the Archaic period that were found to be buried under alluvium amounting to 4m in depth (Bevan 1995).

Each system is dependent upon 'optically pumping' atoms of alkali vapour. It is known that if a magnetic field is applied to a vapour then the valence electrons in the atoms are raised to a higher quantum state. The separation of the energy levels is dependent upon the total intensity of the surrounding magnetic field. This influences the absorption of light from a source of the same material as the vapour. In most modern systems caesium is the alkali of choice as it has only a single isotope and requires little energy to vaporise the element. Absorption occurs in the material as the excited electrons precess about the magnetic field at a specific frequency. In the case of caesium this is about 175kHz. In practice an external field is applied to raise the electrons to this precession state and the intensity of the absorbed light is modulated at the frequency of the precession. The measure of the precession frequency is related to the strength of the magnetic field.

An important contribution in the development of the instruments came when Helmut Becker (1995) charted the use of a caesium vapour magnetometer configured in gradiometer mode. The original set up that Becker employed was a single gradiometer on a cart and the traverse distance was measured using an optoelectronic determined pulse. This system was thought to have a sensitivity of 0.1nT and, due to limitations in the logging system, values were sampled at 0.5m along each line. During 1994 his system was redesigned in collaboration with Picodas using Scintrex CS2 sensors. In this case a sensitivity of picoTesla (1pT = 0.001nT) was claimed at 10Hz cycles. Additional filters allowed the use of the instrument directly under power lines and near electric railways. The data sampling was achieved at 0.1s time intervals and the data were re-sampled at 0.1m distance intervals. In a comparison between his two systems he illustrated that some individual posts within the timber palisades were only found in the new data. This is important because low level responses are evidently of great interest in certain types of sites especially where enhanced soil is barely above the local background. This is obviously true in areas where accumulations of magnetite are formed via the decomposition of organic material by magnetic bacteria (Fassbinder *et al.* 1990). Becker succinctly summarises his conclusions:

> The high sensitivity and spatial resolution of the picotesla magnetometer will cover almost the whole range of the wood-earth archaeology in humid areas with biogenic magnetite which documents most of the archaeological heritage of central Europe.
>
> (Becker 1995, p.228)

Electromagnetic techniques for archaeology

The next four techniques can be regarded as electromagnetic (EM) detection devices. This is because they work on similar principles but the frequencies that they operate vary from a few kHz up to a GHz. They are particularly versatile in that depending on the parameters that are chosen, primarily the frequency that the instruments operate at, it is possible to measure either the electrical or magnetic properties of the soil. However, it is sometimes difficult to predict exactly which property of the soils will be measured; in the 1960s an instrument called the Soil Conductivity Meter (known better as the 'banjo'), was used for investigations at Cadbury Castle. The responses 'were surprisingly clear and positive' (Howell 1966, p.22). However, later tests over ditches known to have poorly enhanced magnetic fills produced little or no response. In due course it was found that the 'conductivity' meter was actually responding to the magnetic properties of the soil (Tite and Mullins 1970).

Electromagnetic survey

All EM surveys make use of the response of the ground to the propagation of EM waves. Essentially most EM systems have separate transmitters and receivers, although some have the distance between them fixed. Put simply the explanation for these instruments is that the transmitter coil induces an alternating magnetic field which creates electrical currents in the ground which are scaled to the conductivity of the ground at that point. These currents produce a secondary magnetic field that is measured by the receiver coil. The magnetic field is related to the conductivity of the soil in the environs of the instrument.

What can EM systems detect?

As these systems are particularly sensitive to the conductivity (the reciprocal of resistivity) of the ground, they are used to locate similar features to those investigated by resistance survey. EM surveys can be used for mapping the remnants of mounds, tracing in-filled fortifications, locating buried stone structures or rubble, pits, and metallic artifacts. The major plus for EM systems of this type is that the current is induced in the ground without a physical contact. This lack of contact means that EM systems can be very useful on land where connection with the surface is variable or where the contact resistance is very high, such as sand.

Principles

The most important aspect of modern EM systems, which operate at a few kHz, is the ability to provide a measure of both the magnetic susceptibility and the electrical component of the soil. This can be calculated because while the rate of change of the magnetic field measured in the receiver is proportional to the conductivity, the magnetic signal is related to the strength of the magnetic properties of the soil. The conductivity is expressed as Siemens per metre (normally mS/m), while the magnetic response is usually reported as parts per thousand (ppt) of the primary magnetic field or the dimensionless units of magnetic susceptibility (SI). Frequently the concepts of in-phase and quadrature signals are seen in the literature and these relate to the phase in the receiver coil of the alternating field. They can be equated with magnetic susceptibility and conductivity respectively. Sometimes the manufacturers of the instruments prefer to see the in-phase component as a metal detector mode rather than susceptibility.

The latest version of the *Geonics* EM38 instrument (**colour plate 2**) allows both values to be measured simultaneously. A further factor is orientation of the coils, either vertically or horizontally, which relates to the penetration depth. Some instruments are capable of simultaneously taking measurements at a number of different frequencies, but their potential in archaeology is still to be proved.

One of the reasons why EM instruments are used in drier climates is that they do not require contact with the ground and because they perform better than electrical resistance techniques on sites with a dry surface. This means that EM systems can be used in the height of summer, over tarmac surfaces or in areas with friable or poor quality surfaces; in other words, in survey areas where electrical resistivity techniques often fail.

EM instruments are frequently used for archaeological work on the European continent (Tabbagh, 1986), and while it is possible that in the future they may find a greater role elsewhere, there has been little reported increase in use in the past decade. Perhaps a stumbling block is that the physics behind these instruments is very complex and the discussion here is highly simplistic. There are many factors that will affect the results from these instruments, not least the orientation of the coils.

Depending on the various parameters that can be changed in an EM survey, the archaeological features that can be detected are similar to those for resistance survey and susceptibility.

Magnetic susceptibility survey

The actions of humans can have a profound affect on the value of the soil magnetic susceptibility (χ). In particular, in the vicinity of a buried archaeological 'site', the value of the topsoil susceptibility is often 'enhanced' by comparison to the local background. The enhancement is usually the result of destruction of the underlying features or strata by ploughing, animal or natural activities. As a measure of the variability of the topsoil susceptibility the technique will produce an additional part of the jigsaw when attempting to work out the significance or variability of the buried archaeology.

What can magnetic susceptibility surveys detect?

Magnetic susceptibility surveys can delimit spreads of enhanced material in the topsoil. Large or long-lived settlement sites tend to produce significant anomalous areas. Conversely, any short-lived or small sites tend to be difficult to detect. An additional use of χ is for predicting the success, or otherwise, of a magnetometer survey at a site. In this case samples from both the topsoil and subsoil should be measured. An estimation of the likely magnetometer response can then be modeled.

As stated above, gradiometers, which are passive instruments, respond to *change* in χ. However, there are a number of cases when the spread of χ, or the absolute value, may be important. This has been found to be particularly true in the analysis of landscapes or niches in environmental studies (Thompson and Oldfield 1986). This use can be extended to include zones of significant archaeological 'activity' that produce an increase in soil magnetic properties but leave little or no physical imprint. The absolute value of χ may also be the only efficient way to detect enhanced but homogeneous spreads of χ; it is believed that a gradiometer would only 'see' the edges of such a spread though we are not aware of any published examples of this.

Principles

The definition of magnetic susceptibility is the ability of a material to become temporarily magnetised. For an absolute value of this induced magnetism an 'active' instrument must be used. The induced magnetisation is a product of both the susceptibility of the material being investigated and the field that is initially applied. As noted above, χ is dependent upon the type and concentration of the magnetic minerals that are present.

Absolute measurement of magnetic susceptibility can be carried out by one of two methods using the favoured commercial system manufactured by Bartington. Common to both methods, which are based on AC induction, is a meter to measure the susceptibility, with switchable sensors that operate at varying frequency, but in the order of a few kHz.

Firstly, there is a field instrument, which allows rapid measurement of large areas. While there are a number of sensors available, the single coil MS2D is the largest and most frequently used (**13**). The system generates an alternating magnetic field of 80A/m in the coil. A volume of soil or other material introduced to this field will affect the frequency in proportion to its magnetic susceptibility. The value of this measurement can be read off the LCD of the portable meter in either dimensionless SI units or the older cgs units. There are no integral logging facilities with this machine, but basic data capture can be undertaken with a hand-held computer. The disadvantage of this method is the poor penetration of the signal, which is a function of the coil sizes that are used. Evidently, while this could be circumvented by judicious use of field sensors, the most frequently used instruments are capable of investigating only the top *c.*10cm of the earth. At first sight this may be a drawback as the majority of buried archaeology lies beneath this depth, but this can be an advantage in that the topsoil is an undervalued archaeological resource. Thus measuring changes confined to this narrow layer may result in mapping former activities that cannot be traced beneath the topsoil.

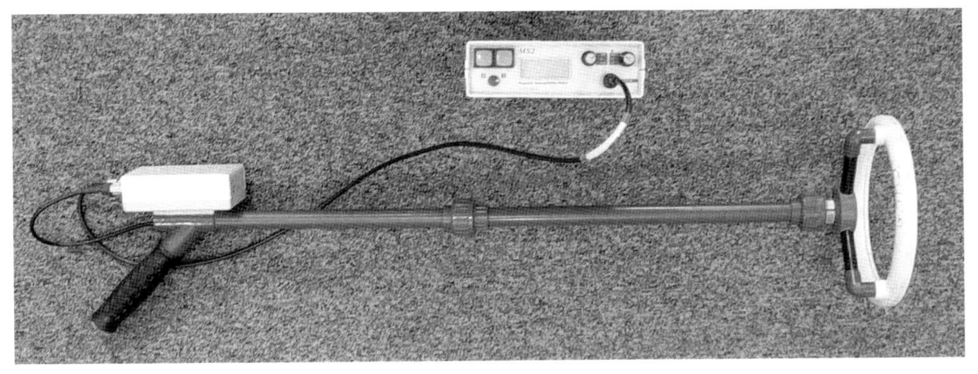

13 *A Bartington Coil. This is used for measuring the magnetic susceptibility of soil in the field. The sensor is the MS2D*

Secondly, laboratory measurements can be made on standard volume or weight of soil. The laboratory assessment is often regarded as a preferred method as it allows a better estimation of the soil's susceptibility as all materials, such as stones and foreign bodies over a certain size (normally 2mm), are sieved out of each sample. The samples are also dried. The additional value of this system is that the laboratory sensors can operate at differing frequencies, which allows an assessment of the magnetic minerals in the material. In the case of the MS2B system the frequencies are 0.46 and 4.6kHz and this allows an estimate of the viscosity of the soil sample (Dearing 1999). The viscosity is the ability of the magnetic grains to align with an external field. Large grain material is said to have a high viscosity as it takes a long time to align with the magnetic field. Smaller grained material aligns much more quickly and this is referred to as low viscosity. In general work, e.g. heating and cooling of the iron in the soil, tends to reduce the magnetic grain size to the extent that this parameter can be characteristic of the use of the area in antiquity (Thompson and Oldfield 1986).

However, the downside of laboratory measurement of magnetic susceptibility is the time it takes to prepare the samples. If analysis of the soil's physical and chemical properties (such as total P, particle size analysis or loss-on-ignition) are to be undertaken, then laboratory analysis of susceptibility becomes feasible.

Metal detectors

Metal detectors, which are specialised EM instruments, are one of the most frequently used tools of the would-be-archaeologist in search of artefacts. Whilst the ethics of their use are not under scrutiny here, it can be stated that the retrieval of artefacts, no matter how valuable, is rarely the main aim of archaeological work. However, when metal detectors are used within a structured research design then there are clear benefits. This is particularly true when looking for burials where the level of geophysical signal from a grave is often very small – under such circumstances a detectorist is more likely to locate metal grave goods or furniture, than the actual grave-cut.

The most popular type of metal detectors are Very Low Frequency (VLF) instruments. They comprise an outer transmitter loop and an inner receiver. A current in the transmitter generates a magnetic field that pulses into the ground. If there is conductive material below the sensor then a weak magnetic field is generated. The strength of the signal is related to the depth of the material and a phase shift between the frequencies of the two coils can be used to discriminate between different materials.

Potentially of greater interest to archaeologists are Pulse Induction Meters (PIM). In PIM instruments, as opposed to standard metal detectors, two coils are also used, but during transmission of the pulsed electric field the receiver is switched off. However, it is switched on about 50 microseconds later in order to detect any responses that return through the ground. By doing this the receiver is unaffected by the transmitted pulse. The responses can comprise electric currents generated by the transmitted field in buried metal or viscous magnetic decay in soil. By momentarily turning off the receiver, the risk of interference between the transmitter and the receiver is avoided. Modern instruments now use one coil to perform the same functions as the original two.

This PIM is a versatile instrument and the first archaeological testing by Colani and Aitken (1966) concluded that it could be used both as a metal detector and as a 'pit-seeker'. This can be achieved by judicious management of transmitter pulse characteristics, sampling delay times and by changing the size and arrangement of the coils. It is of interest that the magnetic property of soil features that is measured is the magnetic viscosity and, with some major caveats, that may be regarded as proportional to the soil susceptibility. On many sites it is therefore possible to conduct area surveys with a PIM rather than conducting a magnetometer survey. Given the fact that the PIM is an active instrument it can potentially give more information than a magnetometer on spreads of enhanced, but similar, susceptibility; generally a magnetometer responds to the change in susceptibility. There are many reasons why this instrument has fallen out of favour, and they include potential depth limitations, but the time is probably right for a revisit to this underused technique.

Ground-penetrating radar

In GPR surveys VHF radio pulses, in the range of *c.*10MHz to 1,000MHz, are directed downwards into the earth from a transmitting antenna. When they meet discontinuities, or surfaces such as floor levels, some of these pulses are reflected back to a receiving antenna; others continue down to be reflected back from other buried features. By measuring the time for the reflection to return, it is possible to estimate the depth of the targets along a vertical section. This initial ability to 'see down' to different depths proved to be a source of great interest to archaeologists who had come to expect 'plan' information from geophysics. As a consequence, GPR has been one of the growth areas for research in archaeological geophysics in the last 25 years. The technique is being used more frequently in archaeology and, as we shall see in the next chapter, several systems are now available. Despite the collection process, down a traversed section, the primary advantage of GPR is its ability, when more than one section is investigated, to provide a three-dimensional view of a buried site. To prepare such a map requires data to be collected along systematic and often closely-spaced transects which are then welded together to form a cube and re-sampled horizontally. The outcome is then a series of subsurface 'plans' at increasing depth, known as 'time slices'.

What can GPR detect?

This technique is particularly good at mapping brick and stone foundations or when depth information is required. Although there are many factors to consider, such as the depth of burial, surface conditions and the nature of the potential targets, it can be said generally that GPR works best where there are good contrasts in electrical resistance.

It is of more than passing interest to note that in Britain GPR is frequently used in built-up areas, whereas in other countries it is often primarily employed for detecting archaeology in rural locations. In Britain it has been argued that this technique is best used in small-scale evaluations over deeply stratified areas where traditional prospecting techniques do not perform as well, e.g. urban sites (Stove and Addyman 1989). This is not because the technique performs particularly well in urban areas, rather that it performs better than any other technique. As a result of the many variables there is often more interaction needed between the radar specialist and the archaeologist, especially at the level of data interpretation (Atkin and Milligan 1992).

While GPR clearly has its place in urban archaeology, where few other techniques will work, it also has its limitations on such sites. Invariably, these are sites with very complicated stratigraphy that would be difficult to understand by trenching, let alone by a non-invasive technique which is influenced by all subsequent activity.

In Britain, GPR is rarely, but increasingly, used on green field sites. There are two main reasons for this. Firstly, other techniques such as gradiometry and resistance survey are suitable and they are quicker and cheaper. Secondly, the high clay content of many British soils limits the effectiveness of GPR, with a rapid attenuation of the signal and a consequential inability to record data to an adequate depth. However, in many ways 'quieter' green field sites produce clearer GPR results, especially over buildings. GPR surveys at Wroxeter Roman city (Nishimura and Goodman 2000), for example, have produced excellent results. While the speed and cost of such surveys prevent its use as a general prospecting tool, it can be very useful as a follow-up, targeted investigation.

Principles

GPR can be regarded as part of the EM suite of techniques that is available for prospecting. A short pulse of energy is emitted from a transmitter antenna and echoes return from interfaces with differing dielectric constants. The travel times are recorded by a receiver antenna and converted into depth measurements, giving a geo-electric depth section.

The penetration of the GPR system is largely dependent upon the frequency that the transmitter emits the pulse of energy. Most commonly the *central* value for the pulse is within the range 200-500MHz. However, antennas within the range 80MHz to 1GHz (=1000MHz) are used for archaeological purposes. These systems are not tunable and are manufactured to emit energy at a specific peak,

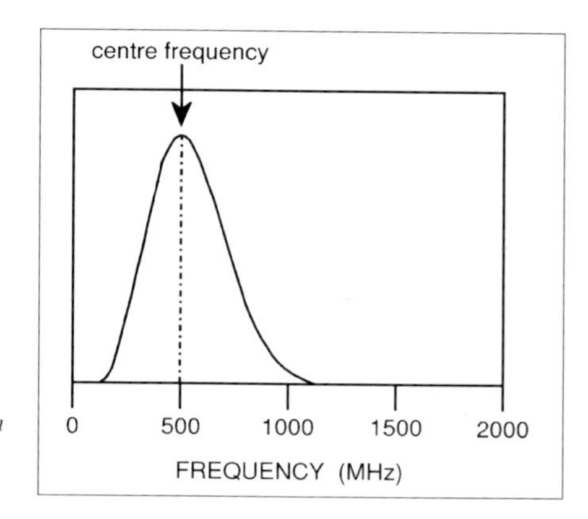

14 *This diagram shows the spread of frequencies from an antenna with a central or notional 500Mhz frequency.* After Conyers and Goodman (1997)

although they actually produce a spectrum of energy as shown in figure **14**. The higher the centre frequency of the antenna, the smaller its physical size and the poorer the penetration of the energy. By way of contrast the lower frequency antennae are so large that they often have to be towed by a motorised vehicle. According to Conyers and Goodman (1997) important factors in choosing which antenna is to be used include:

- Electrical and magnetic properties of the host environment of the site.
- Depth of study.
- Size and dimensions of the archaeological features to be resolved.
- Site access.
- Presence of possible external electrical interference of the same frequency as the radar waves that will be generated in the ground.

The electrical properties of the soil are the dominant factors that have to be taken into account when trying to assess the likelihood of 'success' at a site. Essentially the energy will be dissipated in material of high conductivity, so much so that even thin lenses of clay may render a survey unfeasible. Water with a high salinity, such as seawater, will also cause loss of signal. While this is clearly the case for GPR, this limitation is true for all EM systems.

Resolution of a GPR survey is really dependent upon the wavelength (λ) of the energy that is pulsed into the ground. High frequencies have shorter wavelengths and will be able to resolve anomalies that result from smaller features, but the higher energy dissipation will mean that the depth of penetration into the ground will drop. But how small an object can be defined? While theoretical determinations suggest that a vertical resolution of 0.25λ, more pragmatic reasoning suggests that at least 0.5λ separation must be achieved and 1λ is preferable. To give an idea of how different frequencies would fare, a 500MHz centre frequency antenna would have a wavelength in air of *c.*0.6m, while 100MHz would produce a 3m wavelength. However, when the Relative Dielectric Permittivity (RDP), which is the ability of the substance to store and allow passage of electromagnetic energy when a field is applied, is taken into consideration then the wavelength and the velocity can decrease considerably (Conyers and Goodman 1997). Alongside this decrease is, of course, a better relative resolution. Typical agricultural soils have RDPs that will reduce the wavelengths to half or less than those quoted above. Consequently archaeological use of the technique becomes realistic. The resolution of the system is, however, further complicated by the fact that energy is difficult to focus to a point. At any period in time the GPR is not looking straight down into the earth, but at an ovalised footprint that increases with depth (**15**).

The reflections which equate with alterations in RDP may respond to the changes at the interface between strata or materials. High, or strong, amplitude reflections will be produced when the change in velocity at boundaries is large. Therefore the greater the change in velocity of the transmitted pulse, the greater the likelihood that the causative body can be detected. The technique can potentially differentiate between refilled pits and ditches, graves, buried paths and roadways, air-filled chambers, and metal, wood or stone artifacts. It must be noted that the variation in RDP is not necessarily the result of a physical change, but may simply reflect subtle changes in composition that are the result of moisture variation, etc.

To make the best use of the GPR information with regard to the depth of a reflector it is imperative that a good estimation is made of the velocity of the energy through the ground. This is very difficult to do since the velocity changes as the energy passes through materials of different RDP. A compromise has to be reached and a number of different methods have been used in archaeological work.

Probably the best way to estimate the velocity is by direct observation of a reflected wave. This is normally achieved by knowledge of a wall or a similar structure at a particular depth. This test is often undertaken adjacent to the edge of an archaeological trench where the distance to the depth of the feature can be measured. If an open section is available, but no suitable feature exists, then it may be possible to hammer a metal bar into the section at a particular depth. This will be instantly recognised as metal will reflect the energy immediately, giving a characteristic response.

An alternative estimation can be achieved by assessing the time taken to transmit the pulse between two antennae. Often this is in the form of a Common Mid-Point (CMP) analysis where the two antennae are placed on the ground and moved at set times away from a common mid-point. The GPR system will measure a signal directly through the air, the immediate air-ground interface and from reflections beneath the ground. It is the latter that can be used to estimate the velocity of the Earth at that point. There can be some problems with this estimation as the value that will be measured is likely to reflect only the near surface layers and may have little to say about the velocity through deeper archaeological strata.

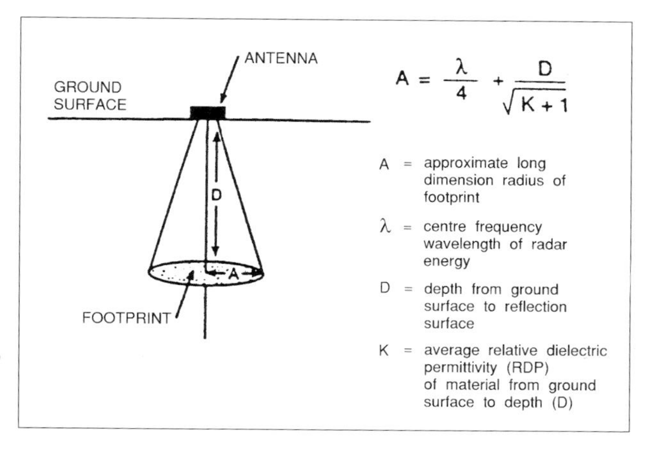

15 *The beam from a GPR antenna does not maintain a collimated form but spreads out with depth. This is a predictable 'footprint' that helps to understand both the resolution of the system and the form of response generated from buried material. In particular this illustrates why a feature can be seen even though the antenna is not directly above it.* After Conyers and Goodman (1997)

Graphical methods are also available in some manufacturers' software that allow a best-fit and laboratory-based assessment, although the latter is an infrequent scenario. If all else fails then there is little option but to select a value from tables of common materials. While this will allow a pseudo-depth, it is likely to produce no more than a relative scale.

A host of processing algorithms are available for GPR data, including corrections for reducing unwanted defects in the data. There is no 'correct' order for each step, but the wrong order may 'blur' an otherwise clear image. One analytical step that is potentially important is that of 'migration'. Migrating the data is an attempt to get rid of the tails of the hyperbola that are part of the form of response associated with point sources, but really are an artefact of shape of the transmission beam (**16**). In theory this should allow a clearer image from small features, as well as creating sharper time slices.

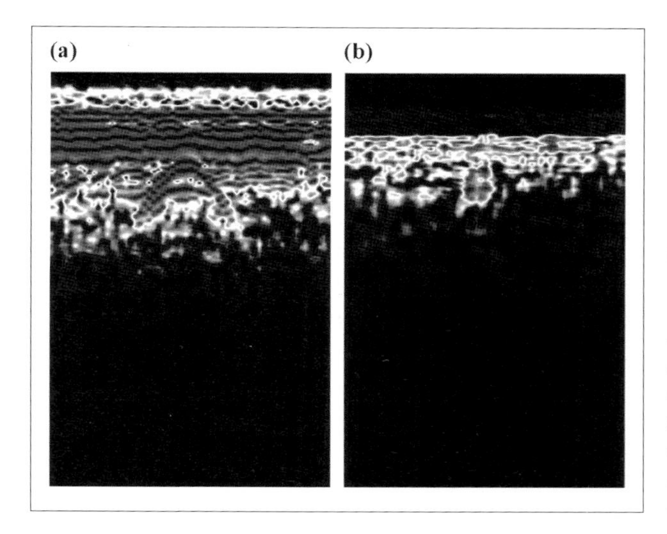

16 *Migration of GPR data. a) This shows a classic hyperbolic reflection from a point source. The 'tail' is due to the geometry of the GPR beam. b) The data after they have been migrated. The 'tail' has disappeared and the response is now closer in size to the origin of the response. Time slicing after migration is likely to produce a truer image due to the reduction of the 'tail'*

Other detection techniques

The following brief descriptions relate to techniques or methods that are infrequently used in a primary capacity in archaeological fieldwork. That is not to say that they are in anyway inferior to those described above, but that their role is often one of support, to validate interpretations that are drawn from other investigations.

Seismic

Seismic survey covers a group of techniques that are extensively used in non-archaeological geophysics. During a seismic survey, artificially generated seismic (or 'shock') waves propagate through the subsurface (**colour plate 3**). The time taken for the waves to return to the surface, by both reflection and refraction at boundaries with differing reflection coefficients, is recorded. The travel times can be converted into depth values giving a vertical section.

While seismic *reflection* has been used for the detection of tombs, it has several limitations. In most archaeological cases the soil layer that is under investigation is relatively thin and may be beyond the resolution of the method. Interpretation of seismic reflection can be extremely difficult when studying boundaries of complex geometry.

As a rule seismic *refraction* surveys are better suited to archaeological prospecting as they can give detailed information about a small area, and the data collection and processing is relatively simple. Refracted rays of energy run along significant interfaces while returning some of the rays back to the surface of the ground. This technique can be undertaken in large-scale surveys, such as the investigation of landscapes and considerably smaller scale investigations of individual features. The source of the energy is usually either a large hammer smashed onto a metal plate on the ground, or an explosive charge fired down a hole. While the latter may give a cleaner and more regular pulse in to the earth, the former may be enhanced by stacking many signals from repeatedly hitting the plate.

For examples of such a survey, see Goulty and Hudson (1994) and Ovenden (1994). Most case studies have concentrated on finding large features or changes in the landscape. An important consideration is that this technique is best suited to conductive soils. As a result, under certain circumstances it may be considered as an alternative to GPR, which does not perform well in wet or saturated ground or clay.

Microgravity

Microgravity surveys were developed to help with civil engineering problems and although the method is not widely used in archaeology, some archaeological case studies have been reported (see Linford 1998; Slepak 1999).

The way that two masses affect one another is described by Newton's law of gravitation. A mass of material, or a cavity, will have a different density to that of the surrounding area and this contrast in density distorts the local gravitational field, giving rise to a gravity anomaly. In order to establish where an anomaly is,

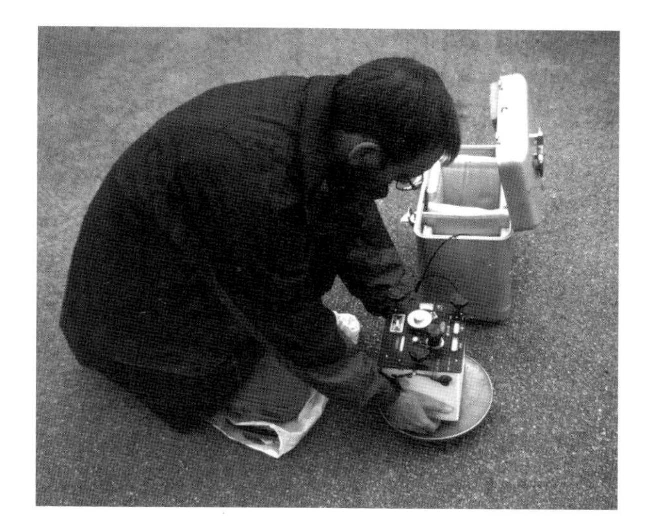

17 *When undertaking a gravity survey the instrument must be accurately levelled. This is a slow process and if a great number of measurements are collected then it is also a repetitive and painful(!) process*

a very careful field procedure must be established using highly sensitive and expensive equipment. The survey method is both time consuming and the post-survey analysis is extensive.

The instruments that are used, such as a LaCoste-Romberg model D gravimeter (**17**), are prone to drift. It is imperative that repeat readings are made at each measurement point and that a base station is frequently remeasured to correct for this drift. Corrections must also be applied for height as well as position on the Earth. The final output is a map of so-called Bouguer anomalies, where the values are normally expressed in either micro-gals or g/cm^3.

On the plus side it is possible to work in a variety of environments, including within buildings, there are also drawbacks due to the slowness of the method. As a result it is not often possible to undertake samples at the collection density that would be required for archaeological investigation. The instruments are also so sensitive that any noise, whether it is passing cars or the 'crashing' of waves onto the sea shore, will reduce the chance of success when attempting to interpret data at the archaeological level.

It is suggested that this technique is used to investigate voids or tunnels where their position is roughly known rather than assessing *terra incognita*.

Induced polarisation, self potential and thermal prospection

The Induced Polarisation (IP) method is similar to the resistivity method and has comparable applications as it makes use of the passage of electrical current through the pore fluids by means of ionic conduction. Induced polarisation is measured by studying the variation of resistivity with the frequency of the transmitted current with the earth acting as a capacitor. As the current is switched off there can exist a small potential difference that can be measured as it decays to zero. The method has been tested on various archaeological sites with little success, although further research is required. Probably the most important recently published paper on IP has shown that the technique was used to find a Bronze Age trackway buried in the Federsee bog (Schleifer *et al.* 2002). This is of great interest as, although IP clearly can detect metallic material (particularly disseminated material as the technique is dependent on surface area), in-filled ditches, and general variations in the topsoil, the fact that wood can be routinely detected opens up application areas where geophysical techniques have been found to be lacking.

The Self Potential (SP) method, also known as Spontaneous Polarisation, is based on surface measurement of natural potential differences resulting from electrochemical reactions in the subsurface. The field procedure is relatively simple, involving the use of two non-polarising electrodes connected via a high impedance millivolt meter. In theory, this method can be used to detect corroding metallic artefacts, building foundations, pits, and underground chambers. However, the technique has not been used extensively and, as a result, its best applications remain unclear.

Buried features can create temperature variations at the Earth's surface that may be measured either using airborne detectors, or ground probes. The effect of buried archaeology on the temperature of the Earth at the surface can be easily recognised on a winter's morning when the frost differentially melts over earthworks. This so-called thermal detection is a continuing area of research, although the data are slow to collect at ground level, and may be difficult to interpret (Bellerby *et al.* 1990).

Dowsing

There are few topics that are more likely to create a debate among a social gathering of archaeologists than dowsing. What can be said is that it has long been practised by a minority of archaeologists and others with an interest in the subject. Unfortunately there appears to be a tendency to detect those features which the practitioner 'feels should be there', and while there are many anecdotes, largely connected with finding water, the case studies tend not to be corroborated by any firm evidence. There are few publications that have dowsing as a core investigation technique and the most cited volume, by Bailey *et al.* (1988), has been seriously questioned and found wanting (van Leusen 1998). The debate regarding this technique, such as it is, is fraught with claim and counterclaim. In our experience dowsing does not work as an archaeological tool. However, our tests are not definitive and unfortunately few dowsers have published their results. The scientific principles, if there are any for dowsing, are not understood and until a set of true scientific experiments can be undertaken then the debate will not move any further forward.

3

METHODS AND INSTRUMENTATION

For many archaeologists the fantasy tool kit would include not only the routine instruments of the trade, such as trowels and tapes, but more exotic items such as an instrument to detect all buried walls, ditches and artefacts. Unfortunately the reality of archaeology is that while it is possible to walk into any local hardware store and get a quality trowel, finding a Universal Detection Device is more of a problem. In fact when it comes to detecting buried archaeological features it should be understood that there is no single instrument that will detect everything. Also, it is important to comprehend that there is no such beast as a 'wall or ditch detector'. Geophysical instruments simply measure various physical differences that can be *interpreted* as being the result of a specific type of archaeological feature.

Whichever technique is chosen there are a number of factors that are common to all:

1. 'Background levels' must be established – that is, the surveyor has to be able to grasp what level of 'noise' is present. The background can vary depending upon a number of factors that have little to do with the buried archaeology such as geology, pedology, topography, agricultural practices, the particular instrument used and even which operator is using the equipment.

2. Whatever phenomenon is surveyed there should be a measurable difference between the 'target' (feature) and the background in which the target is located. This is usually referred to as a contrast and it is this characteristic that creates the measured anomaly, i.e. a zone of readings that are anomalous by comparison to the background.

3. In terms of physical size the target must not be too small for the technique or parameters chosen. If it is then even the most sensitive of instruments will not be able to resolve the response from the feature into a significant signal.

4. The depth of the target is also very important for working out if there will be success in detecting a particular feature. In environmental geophysics a rule of thumb suggests that a target may be found at a depth of up to 10 times its size. While this may be true for the detection of, say, a straight tunnel cut through bedrock, that rule cannot hold in an archaeological context. The main problem is that anthropogenic features rarely conform to set patterns or strict sizes. Most of the methodologies that are used for archaeological prospecting probably only search to a depth of *c.*1m, so the rule of thumb here should be expressed in terms of *assumed depth* rather then multiples of the target size.

Following a brief discussion of each survey method, we will consider some commonly used instruments for resistance, magnetic and radar surveys, and discuss relevant good practice for each technique.

Resistance or resistivity surveys

The earliest instruments used in archaeology have already been referred to in chapter 1, that is the Mega Earth tester, the Martin-Clark meter and the Bradphys instrument (see also Clark 1996). These are rarely, if ever, used in serious work today and will not be considered further here. The domain of area resistance surveys is dominated by instruments from Geoscan Research. For electrical imaging (pseudosection) the systems marketed by Campus and Lund tend to be the most widely used in archaeological work.

Area resistance surveys

Area surveys in Britain are for the most part of the Twin-Probe type, probably due to its comparative ease of use in the field and the interpretation of the data collected. Once set up, the operator only has to concentrate on the two 'mobile' probes that are mounted on a frame.

The great advantage of the first Geoscan Research RM4 instruments, unlike the preceding Bradphys instruments, was their ability to take readings without a need for back-off or balancing controls. Thus, in the late 1970s it became possible to record readings much more quickly because there was little delay in the measurement and display of the resistance value. The RM4 was housed in a watertight, but lightweight box, with a very compact design and a long battery life (**18**). Large area survey became more feasible and there were few problems in working in all weathers, although the resistance values still had to be written down. The appearance of another box of electronics with a membrane keyboard, but no LCD, was the next step in the major period of innovation that has already been described in chapter 1. This box, known as a Geoscan Research DL10 (**19**) allowed automatic logging of readings onto a microchip. It contained a sophisticated menu, hidden on different layers within the system and accessed by listening to a number of beeps,

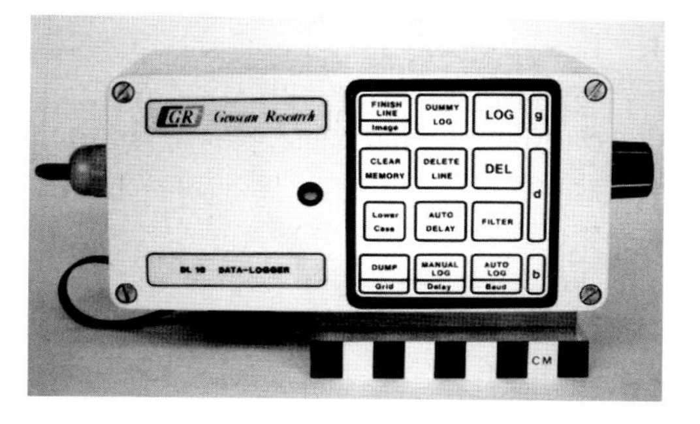

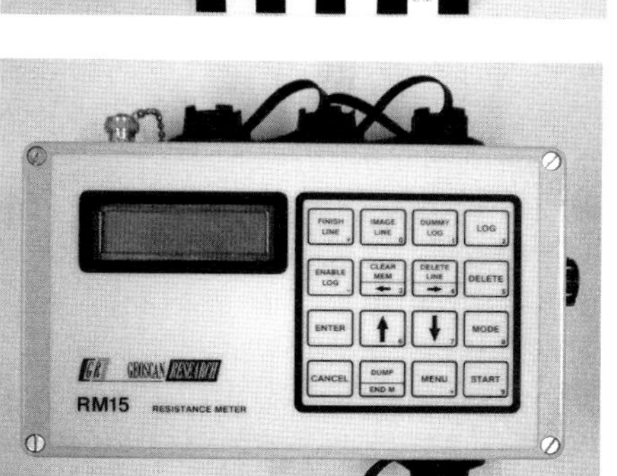

18 (Above left) *A Geoscan Research RM4 resistance meter*

19 (Above right) *A DL10 data logger. This was manufactured as a dedicated logger for the RM4, but could also log data from other manufacturers' instruments*

20 (Left) *The latest resistance meter from Geoscan Research is the RM15 which integrates data logging capabilities*

chirps and warbles that represent grid size, sample interval and so on. While this approach may now seem somewhat Heath-Robinson, the logger was very cleverly designed in that it could be hooked up to a variety of instruments, both magnetic as well as resistance. The manual and subsequent automatic logging of digital data that could then be downloaded directly into a computer heralded a new era in geophysical survey – namely the onset of single operator surveys.

The state of the art resistance instrument, at the start of the twenty-first century, is the Geoscan Research RM15 meter (**20**). With its in-built data logger, capable of storing up to 30,000 readings, and LCD panel, the instrument is a highly powerful and versatile piece of equipment. We will not detail all the features or specifications here but instead point the reader towards the excellent website (www.geoscan-research.co.uk) where this information is readily available. Some of the features, however, deserve mention; the ability to carry out zigzag surveys, and 'finish line' and 'image line' facilities for inserting dummies where the survey area is irregular in shape are all important features. While these are excellent facets that greatly improve the efficiency of surveys, the greatest asset is undoubtedly the auto-logging feature. As the literature says: 'The operator simply inserts the probes of the mobile frame into the ground and the RM15 will determine when the reading has settled, log it, and then give an audible beep to tell the operator to move on to the next reading.' In this mode of operation a 20 x 20m grid can be surveyed at 1m intervals in about 15 minutes. Given that the

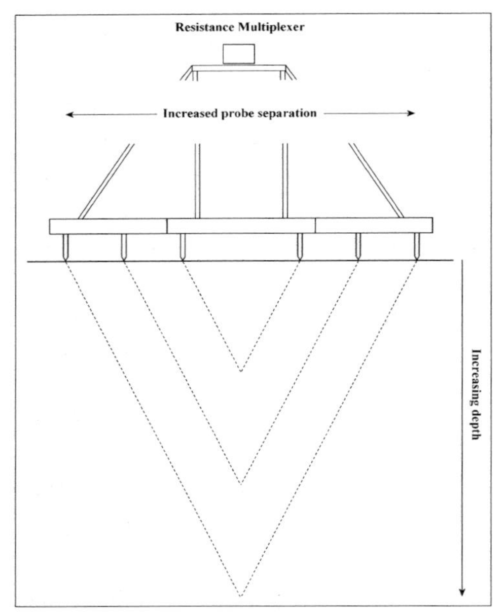

21 *The addition of the MPX15 multiplexer to the RM15 meter allows a variety of readings to be taken at a single survey point. A common use is to collect data at a variety of Twin-Probe separations allowing resistance at differing depths to be mapped. In this case the probes on one side of the mobile frame are current, while those on the other side are potential. The depth of penetration of the current is variable, largely due to the soil conditions at the time of the survey. In most cases the penetration is a maximum of 1 to 1.5 times the separation between the mobile probes. Given this variability it is sometimes best to consider the maps as relative depths, i.e. shallowest through to deepest*

data only take a few seconds to transfer to a computer, the results can be analysed on screen within 20 minutes of starting a survey grid and, if required, printed out in the field almost immediately.

A further innovation, the addition of a Multiplexer MPX15 to an RM15 (**colour plate 4**), provides the professional archaeological geophysicist with an extremely powerful resistance tool. This unit allows what are termed 'Parallel twin' surveys. By bolting on an extra pair of probes onto the moveable frame it becomes possible to log two readings from adjacent traverses before moving on to the next station interval. In this way, for a 20m survey grid, the number of traverses walked reduces from 20 to 10, but 20 lines of data are still collected. Hence the amount of time taken to undertake a survey is greatly reduced. Alternatively, the probes can be mounted closer together to increase the sampling intensity.

But there is another advantage of the MPX15. By adding up to six probes onto the frame and programming the switching system inside the unit, it is possible to investigate a range of measurements. Thus at one station interval a Twin-Probe reading could be taken at 0.5m, 1m and 1.5m probe separations and this effectively provides resistance readings at different depths (**21**). This information can then be viewed as plans of increasing depth or converted into coarse pseudosections. Using the MPX15 system the meter can be programmed to collect information from other probe configurations such as Wenner and Double Dipole concurrently with the variable separation Twin-Probe readings.

Taking resistance measurements with modern resistance equipment is very easy. As the digital recording is automatic, the key to successful data collection is rapidly and accurately locating each measurement position. The most common method is to divide the survey area into convenient blocks, normally 20 x 20m, and data is collected within each block. The most labour intensive approach is to measure each point in with a tape. A quicker way is to mark two parallel sides at 1, 3, 5 . . . 15, 17 and 19m. Lines are strung out between each pair of opposite positions; the lines are 20m in length, made with non-stretch material and have equally spaced marks which indicate measurement positions. The measurement positions start, in the case of 1m samples, at 0.5m and finish at 19.5m; this gives a total of 20 positions, although by measuring on both sides of the line each measurement position can be used twice. Most data collection strategies are based on a variation of this method (**22**). Once data are collected, the multiple grids can be downloaded to a computer and merged together to form an overall map of the resistance values. Modern software allows for the integration of data from many downloads into a single data set.

An excellent example of the adaptability of the MPX15 is shown in figure **23**. Here the instrument's designer has undertaken a multi-layer analysis and covered a 40 x 60m area over a large stone building near the centre of the Roman town at Wroxeter (Walker 2000). He programmed the multiplexer to collect six different Twin-Probe separations, as well as 0.5m Wenner and Double Dipole at each measurement position. While it is remarkable how well the data sets

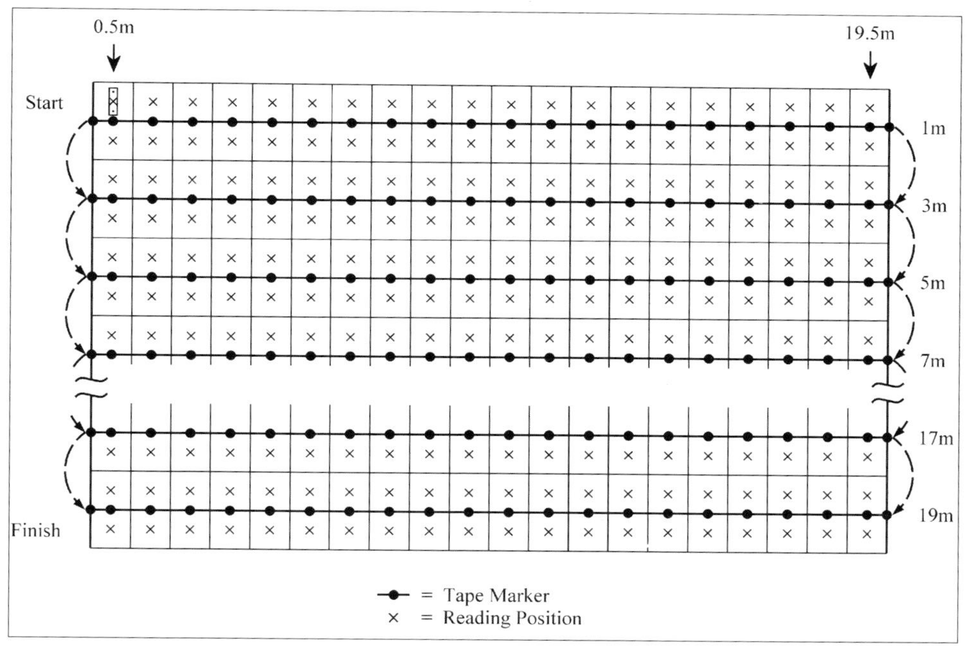

22 *This diagram shows a strategy for collecting resistance data on a 1 x 1m grid using tapes marked at 1m intervals. Note that the first marker is at 0.5m and the last at 19.5m, thereby ensuring that the collection points are symmetrical within the grid. With this strategy two lines of data can be collected using a single line; variations on this strategy can make this method more efficient*

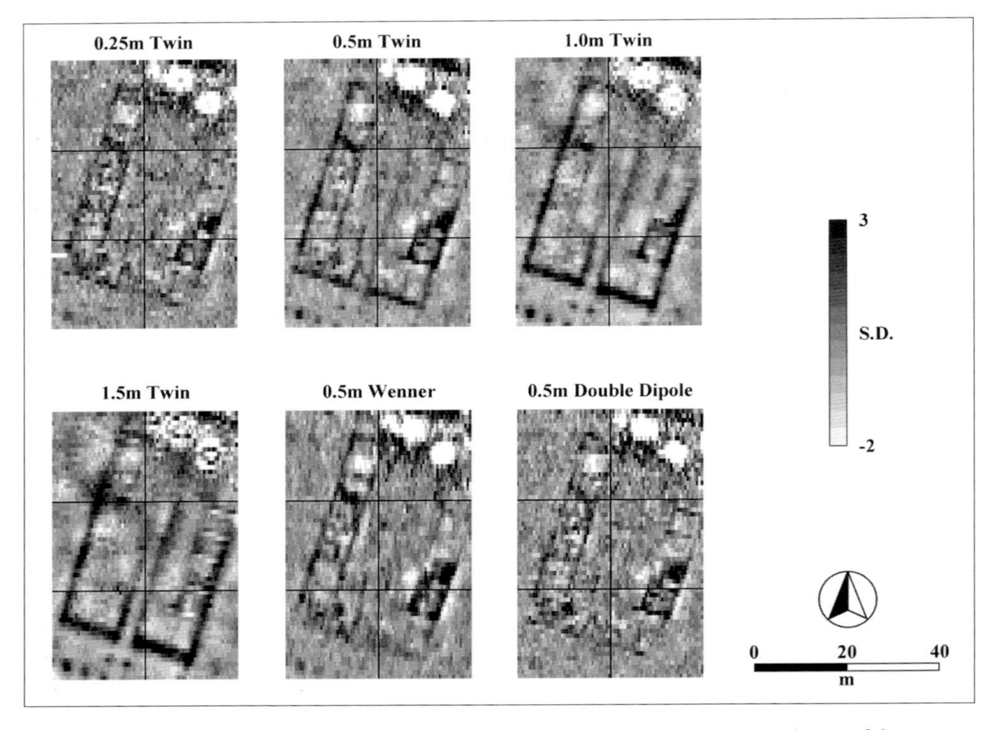

23 *These data, collected by Roger Walker of Geoscan Research at Wroxeter, demonstrate the use of the MPX15 multiplexer in the collection of data using various arrays. The 0.5m Twin-Probe data set is a good compromise between the various options, the data being sharp and not affected by the deeper variation that is seen within the 1.5m data set. Although the Double Dipole and the Wenner data sets appear less sharp, their general compatibility with the Twin-Probe data means that they could be used in areas where remote probes could not be placed far enough away to obey the 30 times rule. See Walker 2000, for additional information on multiplexing*

correlate, there are significant differences between each plot. In broad terms as the probe separation becomes larger less detail is seen within the building and the coarse trend of the underlying geology becomes apparent. The near surface map is dominated by the response from modern 'cattle feeders' (very low resistance) and the appearance of the colonnades at the front of the structure is first seen on the 0.5m data set. This data set provides an excellent validation of the 0.5m Twin-Probe as a standard separation.

Pseudo-section, electrical imaging and tomography

In electrical imaging, pseudo-section work or tomography, the general aim of the work is to provide a vertical section/slice through the ground in order to obtain an estimation of the depth of features or to investigate stratigraphic sequences (e.g. Noel and Xu 1991). These can in theory be combined if a series of vertical slices is put together and analysed in a similar way to radar data to obtain a 3D picture.

The technique is not widely used in archaeology though there are several instances in the academic literature which illustrate the benefits of this type of investigation (e.g. Neighbour *et al.* 2001). However, imaging is widely used in environmental and engineering geophysics when the location of a specific target is suspected.

Field procedure for generating a pseudo-section is relatively straightforward. A number of electrodes, usually 25 or more, are spaced at a set interval along a straight line. The electrodes are connected to a portable computer via a multi-core cable. Software on the computer is programmed to automatically switch between each measurement position which usually takes less than one hour. On irregular or sloping surfaces the height of each probe should also be measured to allow a topographic correction.

Magnetic surveys

The earliest magnetometers used in archaeological geophysics were Proton Magnetometers, which proved successful in the 1950s and 1960s (e.g. Aitken 1974). The principles of their measurement are based around the magnetic properties of protons in water molecules, the variation of which can be related to the absolute values of the total magnetic field. Back in the 1960s, however, the data capture proved to be slow, and we will not discuss proton instruments as now they are virtually unused for archaeological detection. The standard archaeological science text books of Aitken (1974) and Tite (1972) and the recent volume by Clark (1996) more than adequately cover this topic. In Britain it is more usual to use a Fluxgate Magnetometer, where the instrument has been developed for rapid archaeological survey (e.g. Philpot 1973). Another type of instrument that has been used routinely in Europe is the Alkali Vapour or Caesium Magnetometer (alternative names for this instrument are rubidium, optically pumped or optical absorption magnetometers) and it is beginning to gain more favour in Britain.

Fluxgate magnetometers

These instruments use two fluxgate sensors placed vertically above one another, at a set distance apart (either 0.5m or 1m) and measure the vertical component of the Earth's magnetic field. Whichever system or sensor separation is used, it is vitally important to set up the instrument correctly – not only should the sensors be accurately aligned coaxially, but also they should be held as near as possible to the vertical during data collection. The key to the operation of the instrument is the alignment of the two sensors. Field operation with an instrument that has been poorly set up will produce results that are a product of the misalignment of the sensors, rather than the buried features. Although the instruments are relatively

24 (Left) *A Littlemore 1m separation fluxgate gradiometer photographed in 1977. Readings were manually recorded at 1m intervals at the intersection of tapes*

25 (Above) *An early FM series fluxgate gradiometer. They had limited data collection capabilities and the logging sequence was instigated by an external hand-trigger device*

robust, if they are knocked in any way then it is likely that they will need resetting. The most commonly used instruments are also prone to drift with time and this is particularly obvious when the ambient temperature is high or variable. On hot, sunny days the instrument will need more attention than during a cold spell; if a survey is undertaken in a particularly hot environment then it is usually best to collect data early in the morning and late in the afternoon.

Fluxgate instruments often have a noise level of *c*.0.1nT, so very weak changes in magnetic susceptibility can be detected; but, it cannot be stressed too strongly that the set up and use of the instruments is vitally important if one is to achieve measurements at this level. These instruments are highly direction-sensitive and operators must be devoid of ferrous material. The latter includes obvious examples, such as belt buckles and zips on clothing, but also includes material hidden away in the soles of boots or sown into clothes.

As discussed in chapter 1, the earliest gradiometers used in Britain were invariably built by specialist groups or individuals with an interest in archaeological geophysics. The Oxford-based electronics group Littlemore and the larger company Plessey both developed 1m instruments (**24**). These were quickly superseded by the comparatively lightweight Philpot 0.5m instrument which made large-scale surveys far more feasible. The basic limitation of all these instruments, including the basic

Geoscan Research FM9 gradiometer, was the lack of an integral logger. The readings had to be written down manually, linked to an X-Y chart recording system (see Clark 1996); a later option was to hook up to the DL10 data logger that was built for the RM4 resistance meter. Various experiments for automatic logging were carried out using the instruments connected to portable computers via long cables (Kelly *et al.* 1984) and using ultrasonics (Sowerbutts and Mason 1984).

1985 saw the arrival of the Geoscan Research FM18 and FM36 with their internal data loggers, capable of storing 4,000 and 16,000 readings respectively and one-person surveys became possible. A simple hand-held trigger device was used by the operator to log readings as they walked along the survey grid lines (**25**). As each 1m marker was passed the trigger was squeezed and the reading at that point was logged. Although fairly lightweight, extremely rugged and waterproof instruments, the fluxgate sensors do tend to drift and require occasional tweaking during operation. Although this has remained a constant problem with all the fluxgate instruments, the errors can be easily corrected by software.

The next stage in the development of the instrument was the addition of a sample trigger known as an ST1 which was attached to the front of the instrument (**26**). This device allowed the operator to set the grid size, number of readings to be logged per metre and the speed of walking; slow for poor ground conditions, faster when conditions allow. Having marked the grid points on the ground, the operator merely has to walk at a constant speed between the start and finish points and the readings are automatically recorded on a timed basis. The later instruments have an increased size memory chip, allowing more data to be collected at far greater speeds than the hand logger. In fact Geoscan Research, in 2002, added the

26 *An FM36 fluxgate gradiometer with an attached ST1 sample trigger. The ST1 allows various sample densities along a line and these are collected automatically at a set but variable rate*

FM256 to its range and this incorporates the ST1 trigger internally and has a 256,000 reading non-volatile memory (**colour plate 5**). This is coupled with improved times for data downloading and greater battery life.

Single instrument surveys: field method for an FM series magnetometer

Prior to data collection the instrument must be set up within certain tolerances and it is best to switch the instrument on and allow it to warm up for 20 minutes before starting the setting up procedure. In fact this period can usually be used to set out and tie-in the grid. Once the instrument has warmed up, i.e. when both the electrical and mechanical elements have stabilised, then the instrument can be set up to collect data. At this point the operator should be devoid of all unnecessary magnetic items. While experience has shown that spectacles can generally be worn during this period, plastic frames are definitely preferable to metal. The operator should then scan the ground well away from modern ferrous or brick contamination to find a magnetically quiet spot. If the location is devoid of anomalies then, as the magnetometer is lowered to the ground and kept facing the same direction, then the magnetic value should change by only a few nT. The instrument can then be set up via the manufacturer's guidance using both the E/W and N/S alignment controls. There is also a 'balance' control that matches the sensitivity of the two sensors. Once set up, if the instrument is turned around its vertical axis, or it is inverted, the magnetic value should be within 1nT. The instrument should be checked periodically to see if it is still aligned. If it is not then the process should be repeated. How often this is checked is very much dependent upon the field conditions. The check will be most frequent when the ambient temperature is high or changeable, the magnetic background is very low, or if vegetation is high and the instrument is prone to being knocked. Normally data from 2 to 5 grids can be captured before the instrument has to be checked.

Data are usually collected within 10, 20 or 30m grids, although it is possible to programme the instrument for collection over non-square rectangles. It is usual to subdivide each grid in a similar way as for resistance survey, that is pegs or canes are placed at 1, 3 . . . 17, 19m. This time the canes are used as 'sights' for the operator, although some people prefer to place down marked lines similar to those used for resistance survey. In our experience marked lines can be distracting as the data collection is dictated by the regular beep from the attached sample trigger. For a 20m line there are in fact 21 'beeps', the first and last beep coinciding with the edges of the grid. As with resistance survey the data can be collected either zig-zig or parallel. The benefits of the latter are that the raw data is likely to be of better quality. This is especially true when the terrain is sloping as it is difficult to keep the same constant pace both downhill and uphill. Against parallel collection is the slow speed and the fact that most problems associated with zigzag collection can be corrected by software. In reality the choice is up to the person undertaking the survey; the operator should feel comfortable with the strategy and speed chosen. Most operators do a number of 'dummy' runs to synchronise walking and collection speeds.

Until recently, in most evaluation type surveys in Britain, magnetic data has tended to be collected along traverses spaced 1m apart and at sample intervals of 0.5m. This has in part been due to limited processing power and storage capabilities of portable computers and also the slow download times of the pre-FM256 and Grad 601 instruments. Many of the examples selected in this book are from such surveys where we have used the rather coarse sampling intervals. However, as can be seen from the images, excellent results can still be obtained. While most surveys now exploit the improvements in technology and use finer sampling intervals, there are no moves towards closer traverses in routine commercial work, but developments discussed next may bring about such a change.

Dual instrument surveys

In an effort to increase data capture, Geoscan Research are now marketing dual instrument surveys. To survey in this mode it is necessary to place two gradiometers on a rigid frame and link them together thus enabling surveys to be carried out at either greater speed or with increased sampling density (**colour plate 6**). Interleaving can produce effective traverse intervals of 0.5m or 0.25m, which together with high sample intervals, can produce very high resolution surveys, particularly useful in 'research' surveys (**27**). The set-up procedure for each magnetometer on the frame is essentially the same for single instruments. The most obvious difference is that after setting up two instruments and strapping them

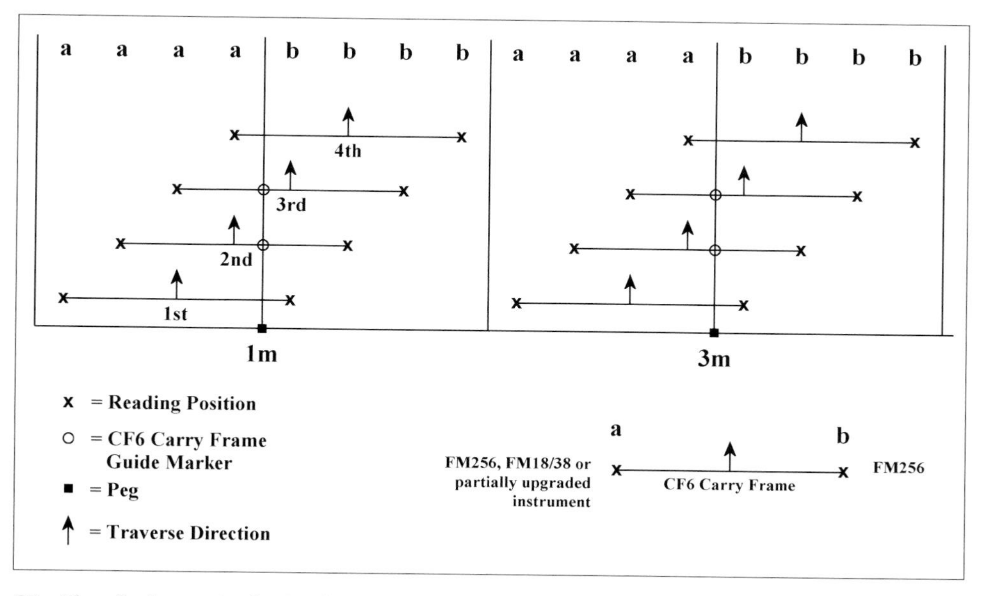

27 *The collection routine for Quad-Density survey using two Geoscan Research FM series instruments attached to a CF6 Carry Frame. At least one of the instruments (the Master) must be an FM256. Using this method a 20 x 20m grid could be sampled at 0.125 x 0.25m intervals.* Diagram courtesy of Geoscan Research

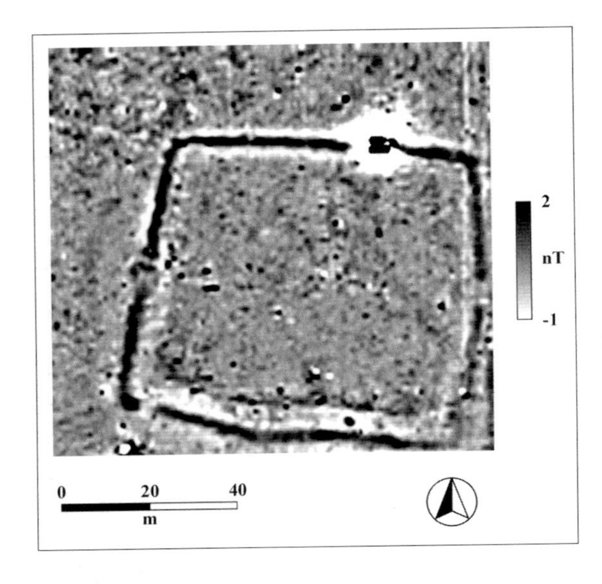

28 *Data collected and processed by Roger Walker using the FM Dual System. The data were collected using two instruments and then merged into a single data set with 4 samples per metre along traverses 1m apart.* Data courtesy of Geoscan Research and the Huddersfield Archaeological Society

onto a frame, one of them is defined as the master and the other the slave. Data collection is controlled by the master.

The collection philosophy for the dual is very similar to the parallel twin mode of operation with the RM15 and multiplexer; two lines of data are collected at once, reducing the total number of walked traverses to half that collected. At present, however, each magnetometer logs its own data so at the end of a survey grid it is necessary to download from each instrument. The data then have to be corrected for any errors before being merged and further processed as required. In this mode of operation a 20m grid can be surveyed in around three minutes at 1 x 0.25m intervals and, in our experience, overall survey speeds can be increased by over 50 per cent using this method. An example of data collected with this system can be seen in figure **28**.

The arrival of a new instrument from the manufacturer Bartington (www.bartington.com), better known in archaeology for their magnetic susceptibility instruments, has seen a return for the first time in 20 years to 1m fluxgate gradiometers. These provide greater sensitivity than their 0.5m counterparts and are theoretically capable of detecting features at greater depths. There are several debatable issues with regard to soil noise effects, connected to the height of the fluxgate above the ground, but given that this can be altered on the Bartington, the issues are somewhat academic.

The instrument is available as a single fluxgate pair (the Grad 601) mounted in a tube like the *Geoscan Research* FM instruments. It boasts a waterproof design, rugged construction and interfaces, and very long battery life. The stability of the fluxgates is excellent and in our experience they are very impressive in the field. They are very lightweight compared to the *Geoscan* instruments. However, the *Bartington* system is effectively modular and is at its most innovative as a dual system fluxgate (the Grad 601-2), which is particularly suited for commercial archaeological survey (**colour plate 7**). The two fluxgate tubes are mounted on a frame which has a central control box that acts as a control panel and data logger. A major advantage of the instrument is that the readings from each fluxgate pair are stored together in one grid file, so there is no need to merge files as is the case with the FM256 dual system.

The set up for the Bartington Grad system is very simple. The instrument has an automatic procedure for balancing the fluxgates that requires minimal operator experience. While the procedure is simple, it is critical to search out a magnetically quiet spot so that both sensors are above a uniform field, i.e. an undisturbed area of at least 2m. This done, the process is automatic apart from pressing buttons and rotating the instrument. Although the system can be set up whilst held in the hand, a non-magnetic tripod is invaluable if the highest accuracy is required. When the system is mounted on a non-magnetic tripod, rotational errors can be reduced to a few tenths of a nanoTesla (**29**).

The frame, central box of electronics and two fluxgate tubes are very lightweight and when carried are supported by a harness that fits over the shoulders. A small weight in a pack slung in the middle of the operator's back, very similar to a small rucksack, counterbalances the instrument. As such the operator is faced with minimal effort to survey large areas, and given the speed of operation due to the dual instrument design, 3-4 ha can be covered in one day, logging readings at 1.0m by 0.25m. An example of data from this system can be seen in figure **30**.

29 *The Bartington 601-2 fluxgate gradiometer has an automatic routine to set up the detectors. This takes between 5 and 10 minutes and for surveys that require the highest sensitivity it is best to undertake this procedure on a non-magnetic tripod*

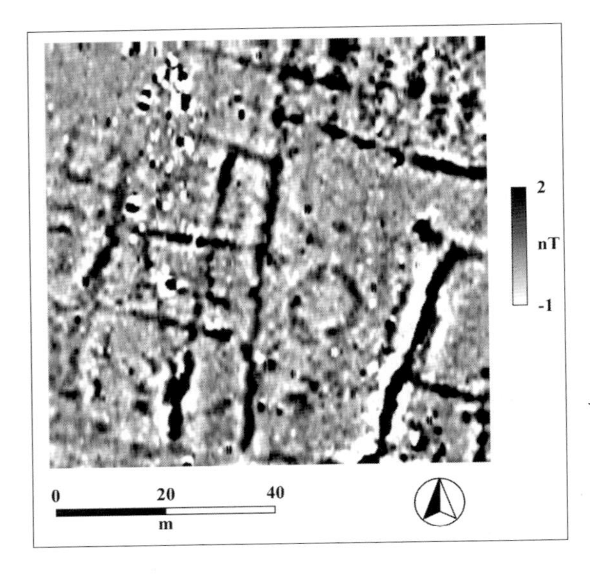

30 *This image shows data collected using the 1m separation Bartington 601-2 fluxgate system. Measurements were taken at 25cm intervals along traverses 1m apart. The survey is at Pinvin (Worcestershire) and was undertaken on behalf of 'The Four Parishes Archaeology Group' (4PAG) who were funded by the Local Heritage Initiative*

Caesium vapour (CV) and other alkali vapour magnetometers

As we have seen in chapter 2, these instruments were first developed in the mid-1960s and they are renowned for very high sensitivity measurements. While their impact in archaeology has been mixed in parts of Europe and beyond, there is evidently great belief in these instruments; in Britain, by contrast, the jury is most definitely out. It is believed that in many areas the concept of even sub-nanoTesla measurement is flawed as the level of the soil background noise is too great. This view is supported by the fact that these instruments have been found to be particularly valuable on soils such as loess, where the natural soil magnetic background is low and the archaeological signals are equally weak. Given these soil factors, it is unlikely that the instruments will find a substantial role in routine archaeological work in Britain. However, there will always be a niche for more specialist requirements, especially when archaeology is buried under alluvium or where timber structures are thought to be present. A good illustration of the use of a hand-held CV instrument (**colour plate 8**) that has been used fairly frequently in Britain can be seen in the data set collected by the Geophysics Section of English Heritage at Stanton Drew (**31**).

Collecting data with a CV instrument

CV instruments automatically store readings on an integral data logger in a similar fashion to the fluxgate instruments. However, in the caesium system data are collected over time, with the operator indicating the start and end points; the instrument takes measurements at a predetermined rate between these two checks. Dependant upon the pace, the operator will collect a variable number of data points along each traverse. The data can then be analysed within commercially available graphical packages, or re-sampled and brought into standard archaeological geophysics software.

In an effort to illustrate the similarity between commercial instruments, we have reproduced data sets collected over a test area using Geoscan Research's FM36 0.5m fluxgate system and Scintrex's Smartmag (SM-4G) caesium vapour (CV) instrument configured as a 0.5m gradient. Data collection was conducted along east-west traverses aligned along the Ordnance Survey grid. The CV data were collected

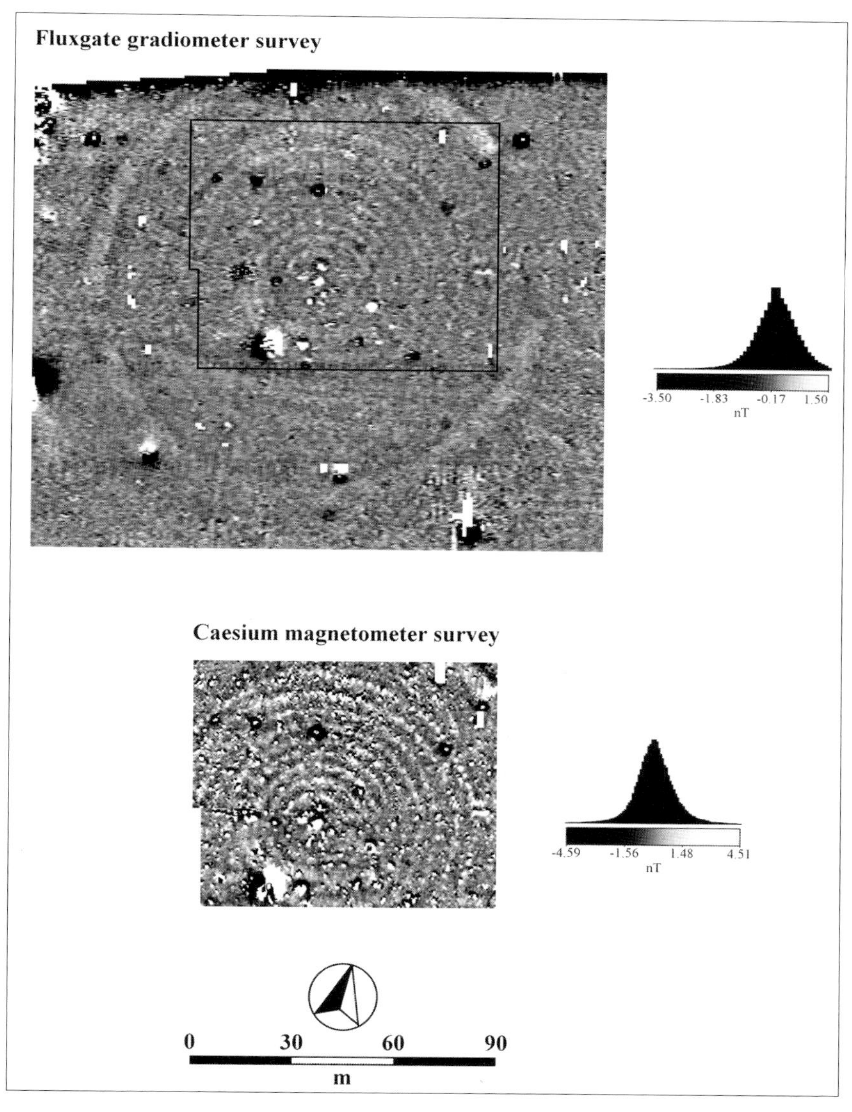

31 *Magnetic data from Stanton Drew collected and processed by English Heritage. The diagram illustrates the similarity between fluxgate and caesium vapour results as well as showing the level of detail that can be achieved when high sampling densities are used. In this case the individual post-holes are resolved thanks to the survey strategy that entailed collecting data at 0.25 x 0.25m for the fluxgate and 0.5 x 0.125m for the caesium instrument. Note the 'vertical' scale is reversed by comparison to the majority of images in this book*

parallel and the FM data zigzag. The latter was set up to collect data on a 0.5 x 0.125m sample within a 10m grid. This instrument, which measures the vertical component of the Earth's magnetic field, was set up on the 0.1nT display. For the former, measuring the gradient of the total magnetic field, a rate was set at 10 samples per second and the instrument has a claimed sensitivity of 0.01nT at this rate.

As can be seen in the diagram, the two sets are highly comparable (**32**). The point made in this assessment is that in both techniques the results have been limited by soil noise rather than the instruments themselves. This means that under similar circumstances there is no benefit from undertaking the survey with a hand-held CV instrument over a fluxgate gradiometer. In use the hand-held CV devices are often somewhat limited in this mode of operation especially when it comes to large area surveys, or when high density data collection is required. To combat this difficulty a number of groups have developed their own custom-built wheeled systems mounted on non-magnetic carts (**colour plate 9**). These wheeled devices, often with a bank of detectors, reduce the signal noise caused by walking and can collect a number of lines of data in one sweep. This is a major advance as reducing the inter-traverse distance is highly beneficial for the detection of small-scale features.

As one can imagine the cost of these systems is considerable, but they have the advantage that the sensors can be mounted in a number of different ways. In the gradiometer configuration a clever and cost-saving implementation is to use only one top, or monitoring, sensor. This is achieved by placing a sensor at such a height that virtually no effect is measured from any near surface features and it is then measuring an approximation of the background for all the sensors. Any minor errors can be corrected in the pre-processing of the data.

A great benefit in using CV instruments is the simplicity of use. As the sensors are not directionally sensitive it means that there are few issues in setting up the equipment. The most important aspect to using these instruments on a wheeled device is the cart itself. The cart is usually built with controls at either end of the device. This symmetry allows the operator to move around the sensors to undertake a zigzag survey without turning the sensors around. This survey mode significantly increases the speed of data collection while keeping the orientation of the sensors common reduces noise. With the wheeled system, as with the hand-held devices, the data are usually over-sampled along the line of each traverse and re-sampled at a fixed distance.

Inevitably there are some issues regarding the use of this form of wheeled device.

- By comparison with the hand-held kit the cart and instrumentation are huge. There can be problems, or at least cost implications, with transporting such large systems. Most of the groups who own such a system have made them modular, so they can deploy only what is needed and can ensure that it can be easily transported to the site.

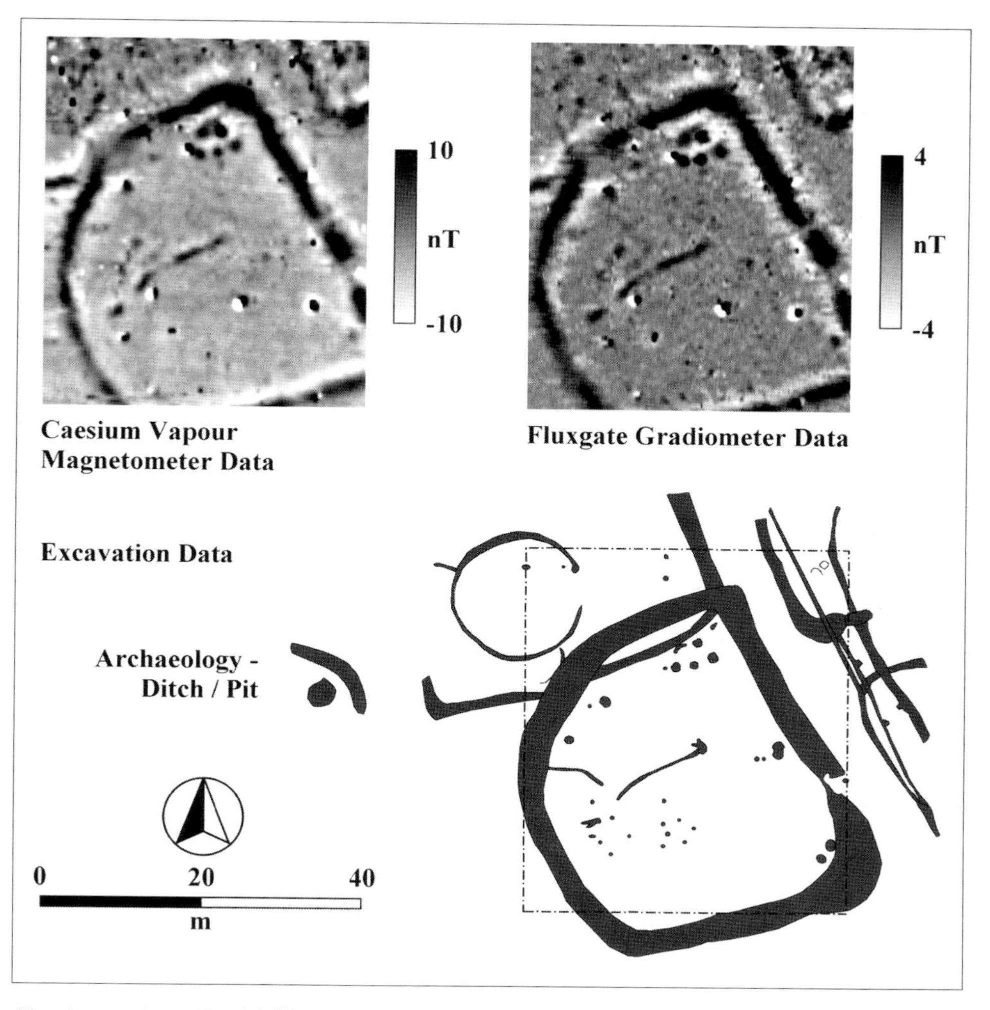

32 *A comparison of hand-held magnetic devices: Caesium vapour and fluxgate gradiometer data sets collected at Grange Park, Northamptonshire. Note the similarity of the two images. Despite the fact that both data sets had traverse intervals of 0.5m, there are still a number of false-positives. Given that they are in both of the data sets it is likely that they are due to material in the topsoil. Some of the smaller features identified by BUFAU during the excavation are not present in either data set*

• To achieve a common top sensor it is normally at least 1.5m above the height of those at the bottom. In areas of overhanging foliage this may prove difficult as the equipment may suffer damage.

• The terrain that such devices can be used on has to be nearer to perfect than for hand-held devices. It is clear that ploughed earth, ridge and furrow, and overgrown vegetation will cause major difficulties when dragging a cart as well as causing excessive fatigue in the operator.

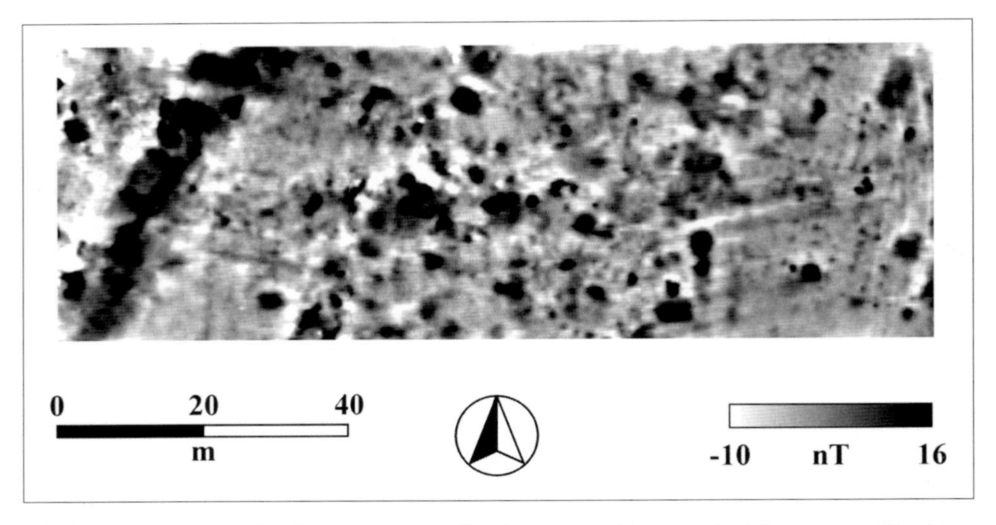

33 *This is an example of caesium vapour data collected using a multi-sensor, wheeled instrument. The data were collected in time mode with a 0.5m sensor distance. The approximate inline mean sampling rate is 10-13cm and the data were re-sampled on a 0.5 x 0.125m raster. For visualisation in a GIS package the image was interpolated (linear) on a regular 0.125m grid. The site is a late Iron Age hillfort with additional occupation in the second to fifth century AD. Excavation and geophysical data are in excellent agreement.* Image left: *fortification ditch dated to Late Neolithic and early Bronze Age, area reused in the La Téne period.* Image middle: *large rectangular anomalies are grubenhäuser from the Late Iron Age and later. A long house, oriented north-south, can be seen between the ditch and the grubenhäuser.* Image right: *post-pits of Germanic houses (second to fifth century AD) dug into the limestone.* The image has been supplied courtesy of ZAMG Archeo Prospection (R), Vienna

• The use of a large cart may prove problematic in small or awkwardly shaped survey areas.

Having noted the problems it is impossible not to admire the data sets that have been collected using these systems (**33**). They have proved immensely valuable in research surveys that cover large archaeological sites and have provided many of the best images in the last decade. In particular, archaeological geophysicists based in Austria and Germany have produced numerous examples of large-scale surveys where image clarity has reached new levels. This has been achieved using closely spaced traverses, often 0.5m or less, and even greater sampling density along each line.

Magnetic susceptibility

It is perhaps surprising that in the past 15 years or so only one commercial system has been readily available for area surveys – the Bartington MS2. It is described in detail on the manufacturer's website (www.bartington.com) and in Clark (1996). Features of the instrument include the ability to toggle between SI and CGS units, single or continuous measurement and a full day's operation on one battery recharge. The system can be hooked up to either a field coil or bench sensor to measure soil samples.

The field coil or search loop, MS2D, is designed specifically for the rapid assessment of magnetic susceptibility in the topsoil and is used widely in archaeological assessments (**colour plate 10**). It is used both to prospect for, and investigate internal details of, archaeological sites (**34**).

Collecting field data with a magnetic susceptibility meter

The field procedure for the Bartington MS2D is very simple. The operator firstly decides whether SI or cgs units are to be used and which level of accuracy is required. Once the instrument has been given sufficient time to warm up the sensor is lifted above the head and the zero button is pressed. This measures the susceptibility (χ) of the air and sets this level as '0' and acts as a standard against which the measurement of the soil χ can be made. This is found by pressing the coil against the soil and pressing the measure button. The χ value is displayed on the console LCD and can be written down or saved on an attached hand-held computer. The weak link in this procedure is the physical coupling of the coil with the soil, especially if the surface is highly undulating. In such circumstances it is often best to take several readings and work out the average. The readings are

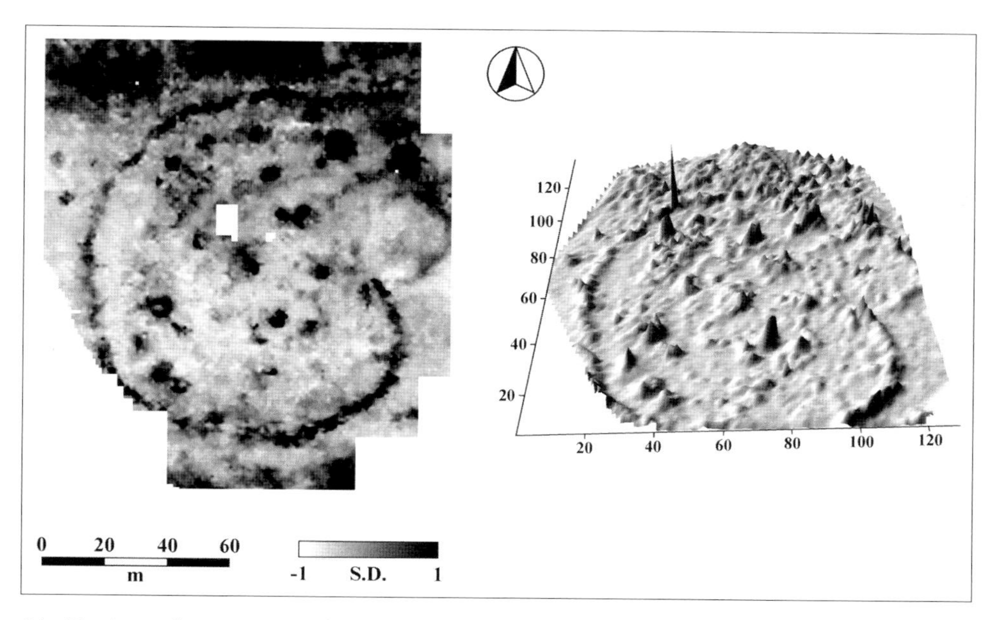

34 *Two images from a magnetic susceptibility survey at King Lobengula's Royal Palace at Old Bulawayo, Zimbabwe. The data were collected on a 1 x 1m grid using a field coil. The site was short lived and, from a geophysical point of view, had the benefit of being burnt to the ground. Excavation by BUFAU indicated that the ditch encircling the dwellings, as indicated by narrow ring of high susceptibility readings, is unsubstantial and has only been found because it was burnt down*

normally taken on a regular grid and when prospecting for archaeology the sample interval is in the range 5-20m. The instrument is prone to drift and it is usually good practice to re-zero the system every few measurements. When there are a few metres between measurement points the instrument can be re-zeroed whilst the operator is walking between locations.

Ground-penetrating radar (GPR)

GPR has been used in archaeology since the mid-1970s (Conyers and Goodman 1997, 18-21) with pioneering work being carried out in New Mexico (Vickers *et al.* 1976), and the States (Bevan and Kenyon, 1975). The take up of this technique was slow, partially as a result of cost and partially due to the lack of digital data. The cost has reduced quite dramatically in the intervening period and digital output is now standard. The latter is a great advantage as less time is spent optimising the output in the field as well as allowing infinite tweaking of the data and alternative display options.

There are a variety of commercial radar systems in use though all operate in a very similar manner. The transmitting and receiving antenna are pulled over the ground and distance markers are added to the data files by means of an automatic odometer (**colour plate 11**) or manually via a hand trigger. In some instances the antenna are mounted on a sledge, in others they are fixed to a wheeled cart.

The large amounts of data that are collected invariably mean that some form of computer is required on site, either as an integral part of the system or as a stand alone PC. Different antennae are attached to the system dependent upon the nature and potential depth of the targets.

Although initially the main advantage of GPR was considered to be the ability to view a vertical section through the ground (**35**), experience has shown that it is far easier to view data as a plan. By collecting many parallel lines of data and merging them together into a block, it is possible to produce a series of time slice, or amplitude, maps (**36**). These sum the data within a selected time or depth range for every traverse and save the data as an XYZ file which can then be displayed as a plan of response at that particular depth range below the surface.

In GPR surveys it is important to note that the quality of the time slice is highly dependent upon the spacing between the traverses. In the past, when GPR data were only analysed by traverse, they were often 5m or more apart. For time slice production the traverses are usually a maximum of 1m apart and this distance is often reduced to 0.5m or 0.25m (**37**).

A number of manufacturers produce GPR systems that are similar in approach (e.g. GSSI, Mala Geoscience and Sensors and Software). For most manufacturers there are two common methods of data acquisition – either 'continuous' or 'step'. The latter captures data at set distances along a traverse. The capture can either be on a timer which tells the operator when the reading has been taken, or taken

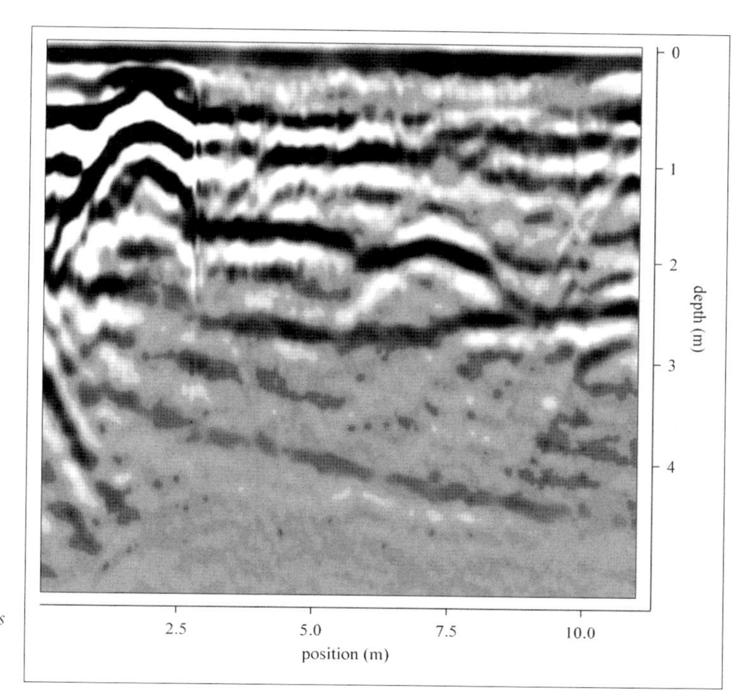

35 *A typical radargram showing reflections from buried features and layers, as well as dipping reflectors from structures beyond each end of the traverse*

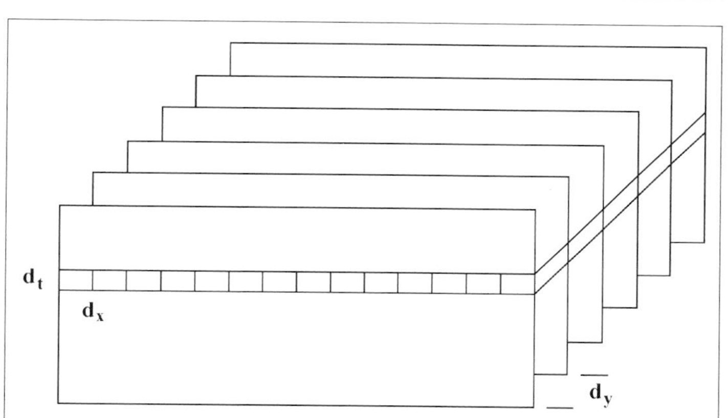

36 *How 'time slices' are generated. Traverses are stacked within a cube and the data resampled horizontally. The thickness of each slice (d_t) can be varied. While this is a common way to analyse GPR data, other data sources, such as Electrical Imaging, can be treated in the same way.* After Conyers and Goodman 1997

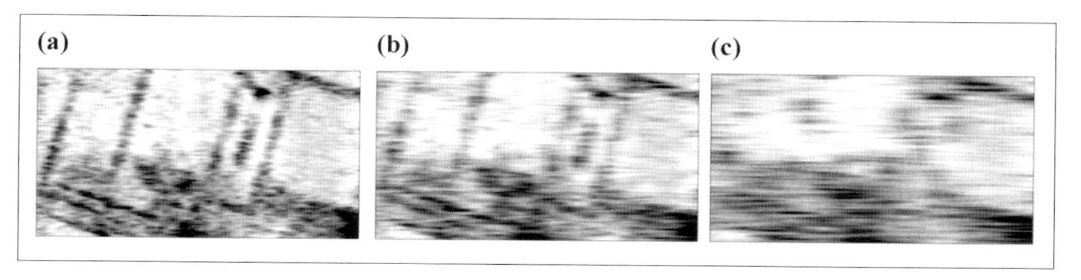

37 *GPR data collected at the Roman town of Carnuntum. This diagram illustrates the need for close traverses if fine detail is required in time slice images. In this diagram a time slice was generated from a) 0.5 x 0.05m, b) 1 x 0.05m and c) 2 x 0.05m data intervals.* Plots courtesy of ZAMG Archeo Prospection (R), Vienna. Additional information regarding sample interval and direction of traverses can be found in Neubauer *et al.* 2002

automatically when attached to a measurement wheel (odometer). For continuous recording the antennae are usually dragged across the surface at walking speed and the operator activates a marker at predefined distances.

Before the data are collected the system must be told what parameters are to be used. Normally there is some form of overall information regarding the job and this will contain common facts such as the centre frequency of the attached antenna. There are a number of other factors that will have to be determined. For example, how much data will be recorded (the 'time window') – it is pointless collecting too much data as it will increase processing time and will give a false impression of the depth of investigation. The number of samples within each trace will define how well the reflected wave is defined. The values that can be used are variable depending upon the system being used although 512 samples/trace is commonly used in archaeological investigations. In an effort to gain better data the traces are often said to be 'stacked'. This involves averaging a set number of traces to increase the signal to noise ratio. Naturally, stacking traces means that data acquisition must be undertaken at a reduced rate. However, if a moderate amount of stacking is undertaken, e.g. eight traces, then a comfortable walking pace can still be maintained. Some GPR users prefer to speed up data collection by stacking very few traces, and they rely on robust post-survey processing (e.g. Neubauer *et al.* 2002). Nearly all data benefit from enhancements, notably in the form of a 'gain'. A gain is effectively a device to boost the signal in a specified part of the trace, notably the later, weaker part. However, most systems allow the data to be collected raw and the gain applied at a later time.

Traditionally, data collection has normally been undertaken with at least two people. One person sits with the computer to activate the data collection and to check that the system is working properly. The other walks backwards, dragging the antennae along a predetermined traverse. If the survey data is being collected in continuous mode then this person will usually have to press a button on the instrument's handle each time a regular marker is passed. As an alternative, an odometer can be attached to the unit and data collecting becomes automatic. A third person may be required to move tapes or markers. Increasingly newer systems are self-contained in that antenna, odometer, battery, control unit and data storage are all built into a wheeled device. The possibility for single-person surveys is obvious.

As the antenna is dragged across the ground the reflected waves are collected, stacked, and displayed in real time. The image will vary depending on the velocity chosen for the depth axis, but like so many processing steps can be refined after the data have been saved.

4

SURVEY LOGISTICS

Armed with a wide variety of geophysical techniques and faced with an enormous range of different archaeological problems that need solving, there are certain fundamental issues which have to be addressed in all survey projects. This chapter will deal with survey logistics: covering basic questions such as the suitability of sites for investigation: base maps; establishing the survey grid; sampling strategies; and how geophysical survey work relates to other elements of archaeological investigation.

What should be understood is that a considerable amount of planning goes into each survey prior to any fieldwork being undertaken. Some of the questions that need asking involve basic issues regarding the end product of the survey. If a clear view of what is to be expected is known from the outset then the chances of perceived, if not genuine, success may be higher. Certainly levels of confidence in the interpretation or conclusions of a report are easier to establish in such circumstances.

The starting point for many projects involves discussions between archaeologists and geophysicists where the aim should be to identify achievable objectives and a research design. The archaeologist may have drawn up a broad scheme, sometimes called a 'brief', or a very detailed plan, which is termed a 'specification'. In a professional situation the geophysicist would produce a method statement to illustrate how the techniques and which methodology would be used to answer the research questions. In real life it is often the case that the objectives are very broad because there may be only the slightest evidence for archaeology prior to the survey. Sometimes the justification for a survey is simply a feeling that there is something in the area and geophysical techniques will be the best way to start the investigation. One area of concern is the small amount of time often allowed for the mobilisation of a survey. Although the implementation of the techniques can be rapid, this is no justification for rushing the strategy. In instances where time is short, or when finances are tight and resources are not available, then an ideal strategy, involving many techniques, may not be possible. Planning a survey often involves compromises, but should never allow a lowering of standards.

A strategy must be developed that produces the best value survey, in terms of archaeological output, in the allocated time available. The actual timing of a project is another consideration and may be one of the determining factors in whether or not a survey should be undertaken at all. For example, should a resistance survey be carried out during the height of summer, or for that matter the middle of winter, when moisture contrasts are often low, simply because that is the only time when the land is available?

If a survey is carried out as part of the planning process it may be that the aspirations of the archaeologist or client are far beyond the likely outcome of the geophysical survey. In such cases it is best for the geophysicist to make known their reservations before getting into the field. The archaeologist generally values a note of caution, but in some circumstances a geophysical survey may be his or her only hope, no matter how slim, to answer a particular problem. A geophysical survey is normally part of a broad research agenda and the geophysical evidence may not answer *the* question but, along with other data from fieldwalking, documentary and cartographic research, etc. may help resolve the query. Occasionally the data simply generate more avenues of enquiry. It is up to the geophysicist to ensure that all the correct questions are asked in order for the survey to be productive.

Survey considerations

From the outset it must be stated that not all sites are suitable for geophysical investigation. If there is tall or dense vegetation and accurate conventional surveying is likely to be a problem, then collecting meaningful geophysical data may be impossible. This scenario can usually be sorted in advance by undertaking a site visit, perhaps by a local contact who can report on field conditions. Emailing digital photographs prior to survey can assist in decision making but it is of limited use photographing or describing the state of vegetation too far in advance since crops can grow extremely quickly. Naturally landowners are disinclined to see crops flattened.

Some factors can be checked out well in advance from standard reference sources, for example the geology, pedology and topography. Perhaps of greatest importance is the nature of the suspected archaeology so that the right technique(s) can be employed. All these factors will lead to an assessment, no matter how qualitative, of how likely the chances of success are at the site in question.

Looking at these general points in greater detail:

Geology and pedology (soils)

The nature of the geology and its proximity to the ground surface will have an effect on any geophysical technique. However, it is difficult to generalise about the suitability of the geology for the 'success' of a survey simply by looking at the parent bedrock. Clearly soil processes and formation will also impact on a survey, and in particular magnetic survey. The following summary attempts to indicate the suitability or otherwise of major parent geologies on magnetic survey:

DRIFT

Glacial till	Good for clay-with flints; variable for boulder clay; good over Jurassic clays and limestones. Igneous cobbles can give spurious pit-type anomalies.
Glaciofluvial	Moderate to poor over upper Jurassic geologies, variable over sandy areas.
Fluvio-glacial	Variable to poor, especially where high water table. Prone to producing 'speckled' datasets with natural anomalies such as bands of gravels or spurious circular patterns.
Coversands	Poor, depending on depth to archaeology and magnetic nature of overburden.
Alluvium	Variable, results depend on depth and nature of target, soil texture, presence of waterlogging and the magnetic nature of overburden.
Marine	Usually poor.

SEDIMENTARY

Most sedimentary parents are suitable for gradiometry. Nearly all limestones give good results, but beware of natural 'cracking'. Cretaceous and Tertiary clays, which include Oxford, Gault and London clays, tend to be poor. Clays, mudstones and shales are not a problem *per se*; however, their heavy, poorly-drained nature made them unattractive to agriculture and settlement. Reasonable results have been had from Carboniferous Grits. Sandstones are variable – Old and New Red Sandstones have tended to give fair results but Greensands are often poor.

METAMORPHIC

Good results have been achieved over slates. The crush or contact-metamorphic geologies encountered in Cornwall give excellent results despite elevated background levels of response. The environs of igneous intrusions, however, are highly problematic.

IGNEOUS

In general, igneous parents are problematic owing to their thermo-remnant magnetism. However, successful surveys have been made over basic igneous parents.

It is acknowledged that the above summary is an over-simplification that can only serve as a rough guide. However, experience of survey over a range of geology and soils is a starting point that geophysicists should always consider. For example, the presence of igneous geology immediately rings alarm bells when magnetic surveys are being planned, as do clayey soils when contemplating GPR surveys. In practice, when considering the effectiveness of a technique, it is the type of the suspected archaeology that is often as important as the geology or pedology.

Research has shown that the inherent magnetic susceptibility of different soils can be a good gauge of the effectiveness of magnetometer surveys. It is now possible in Britain to draw conclusions on possible success rates based upon the results of past surveys on different soils. The English Heritage database is a good starting point for finding information about such surveys.

Vegetation

While it has been shown earlier that vegetation can have a direct effect on the moisture content of the soil and hence any electrical methods, here we address the practical aspects of carrying out a survey, in particular data collection. Any vegetation that impedes the ability of an operator to walk over the ground with the geophysical instruments will cause problems. Naturally the extent of the problem will depend on the technique being used.

Most magnetometers require the instrument to be held in a vertical axis though there is a certain amount of tolerance in each system. However, tall grass or crops which result in the instrument being buffeted as the operator walks along are likely to result in highly increased noise levels and even spurious anomalies. To achieve the best results in these conditions it may be that the automatic collection rate has to be reduced or even manual readings have to be taken. While readings can be hand-logged at stationary points, this is a slow and tedious process only resorted to where conditions suggest that no automatic triggering device will work and yet the project must be completed during a set period. A major problem is that this can only be achieved over small areas and usually the sample density is less than when using automatic data capture. Usually it is better to cut and clear the vegetation rather than survey in dense or overgrown conditions.

The ability to place measurement coils on the surface, push resistance probes into the ground or drag GPR antenna over a surface will also be affected by the vegetation. Wooded areas, bushes, shrubs, bracken and heather all present problems that are very difficult to overcome. Vegetation may have to be cut and removed altogether before the survey can start. In some cases this may not be possible and other ways to investigate the sub-surface may have to be found. This is particularly true on Sites of Special Scientific Interest (SSSI) and where there are commercial crops. If surveying over the latter can be delayed until after harvest, then conditions will be considerably enhanced as well as avoiding needlessly destroying a cash crop. A result of these issues is that the months immediately following harvest are often a peak period for carrying out archaeological geophysics.

Weather

Most instruments are waterproof and designed for relatively rough field conditions. Many contain integral data loggers and their own batteries, so leads and connectors present few problems in light, if not heavy, rain. Torrential rain should be avoided as much for the sake of the surveyor as the equipment! Rugged portable computers that stand up to a certain amount of adverse weather also exist, but in most instances field computers are operated best inside a survey vehicle.

Of the commonly used instruments the one exception is GPR where an external battery is usually required. The design of the equipment is such that connecting leads are often mounted on the antenna in a position where they are exposed to the elements and water penetration can present problems.

Most geophysical instruments will work in both sub-zero and hot temperatures, but as a general rule a minimum of -5 and a maximum of 40°C are perhaps appropriate. For all equipment it is important to read the instrument manual, as operating in weather or temperatures beyond those sanctioned by the manufacturer is a risky and potentially costly affair. Whether the geophysicist can operate efficiently at these extremes is another matter.

Previous work

Given the thousands of surveys that have been undertaken for archaeological purposes it would be surprising if one had not been undertaken on similar terrain to a proposed survey. Naturally, most professional companies are likely to have their own database of results that they can consult. There are, however, a number of databases that are available online that hold important information. Once again English Heritage holds information on a large number of surveys within England, while the North American Database of Archaeological Geophysics can be found at www.cast.uark.edu/nadag/.

Field considerations

Some of the factors noted here are specific to individual techniques, and normally produce localised effects. In particular, many of the problems highlighted below are related to magnetometer survey as this is prone to many complicating factors. Some of these noise sources can, in extreme cases, render survey data incomprehensible. This section contains information from Gaffney *et al.* (2002).

Wire fencing

This is an important consideration in magnetic survey as multi-stranded wire fencing can produce a large distortion in the local magnetic field. As a rule of thumb, magnetic data must be collected at least 1m away for each strand of wire in a fence and the disturbance can usually be detected up to 5m away. Boundaries, even those that do not have a significant ferrous component, can be problematic from the outset as they often dictate the orientation of the survey grid.

Overhead power cables

Contrary to popular belief, the majority of these do not affect the quality of results collected using a gradiometer. A 'shadow' has been found underneath some low power lines, although this has not affected the interpretation of the wider area survey. However, problems may be encountered when trying to set up a magnetic instrument beneath such cables and so this practice should be avoided.

Underground power cables

Underground power cables can produce strong magnetic anomalies that may restrict the usefulness of magnetic prospecting sometimes up to several metres on either side of cables and electrical interference may prohibit resistance survey.

Pylons

Pylons are problematic due to the large mass of ferrous material in their structure. In general 20-30m is the closest the operator can approach with a magnetometer without spurious effects being recorded.

Radio/Cellnet/low and high frequency transmitters

The effects of transmitters are difficult to predict, as the response is dependent on the frequencies at which they operate. In such cases instrument manufacturers may have to be consulted, but even then there may be room for ambiguity. It is often the case that a rapid field assessment may be the only way of evaluating the effect. Experience has shown that, even when a transmission does interfere, there may be some zones where interference may not occur. Interference has been reported when using mobile phones next to GPR systems and low frequency transmitters may disrupt the measurement by resistance meters.

Electrified railways/overhead cables

The ferrous content is the overriding factor in determining whether the instruments will be affected. Passing trains will produce very large magnetic fields which will cause temporary saturation of a gradiometer. Of course, the areas immediately adjacent to railway lines often attract modern and ferrous rubbish and 'plumes' of noise may emanate from these places.

Vehicles

A stationary vehicle can be detected by the gradiometer 20 to 40m away. As with trains, the amount of disturbance is dependent upon the amount of ferrous material in the vehicle. Our experience of surveying next to motorways and major roads is that passing lorries will produce very large, spurious magnetic anomalies. These anomalies are very difficult to suppress as they randomly affect only a small number of measurements as the vehicle passes by.

Buildings

The majority of modern buildings often contain fired brick, magnetic stone, steel reinforced concrete and corrugated iron in some form and often in substantial quantities. All of these building materials produce magnetic fields that are likely to swamp nearby anomalies of archaeological interest. The most problematic structures tend to be warehouses, gas and oil holders, although mobile homes and caravans, including site offices, will present similar problems on a smaller scale. As a consequence, the positioning of temporary site offices should be undertaken after the geophysical survey takes place. However, where buildings are permanent, there is often little that can be done apart from choosing a technique that is not affected by such problems. An additional factor is that the areas immediately adjacent to structures have often been subjected to considerable amounts of landscaping or consolidation that will have a bearing on any standard technique.

Pipelines

Buried ferrous pipelines will have a marked effect upon the local magnetic field. Some of the larger gas and oil pipelines will preclude effective use of a magnetometer at distances of up to 20m either side of the pipe trench. Fortunately the response recorded as a gradiometer passes over a buried pipe is reasonably characteristic (Sowerbutts 1988). However, when displayed as a dot density plot or a greyscale, some pipelines can appear as a line of interrupted high readings/blobs – similar to a line of pits. While a novice may incorrectly interpret a pipe response, a quick view of the data in X-Y format will indicate the true strength of the anomalies associated with the feature. Sometimes a pipe will produce a response 10 or even 100 times that associated with an archaeological feature. Some pipes are protected by cathodic protection (a low electric current is passed through the pipes to prevent corrosion) and this can have a major affect on resistance survey.

Other instruments

If more than one magnetometer is being employed then it must be remembered that an individual instrument will affect the field that a second instrument is attempting to measure if they are within a couple of metres of each another. Second instruments are even more problematic when using active instruments, such as resistance meters; the data that the instruments collect can become very noisy and a measurement value at a stationary point can vary considerably producing unworkable data.

Ground conditions

Modern dumping

Modern material, e.g. lumps of concrete and clinker in the ploughsoil, along with the artificial build-up of ground surfaces (e.g. embankments, consolidation and landscaping) all pose problems, both during data collection and interpretation of the data. Such material can also increase noise due to problems with ground contact, especially when conducting magnetic susceptibility, resistance and GPR surveys. During magnetic surveys the variability of the ground will often be apparent when trying to set up an instrument.

Trees, bushes and shrubs

Generally speaking the presence of trees, bushes and shrubs are tolerable as long as the operator can walk in straight lines between them. Usually this means that they are less than 1m across and that branches are not at head level or lower. If the vegetation is dense then the quality of detailed survey work will be reduced.

Crops, undergrowth and flowerbeds

One of the first considerations must be whether a crop is robust enough to have people walk through it without damaging or reducing the profitability of the crop. Apart from very sensitive crops, there are few that will not survive even intensive data collection. Mature crops are different as the produce is far more likely to be damaged at the late growing stage. Another consideration is whether or not a sufficiently good base grid can be established for data collection when a crop is tall or dense. Furthermore, when undertaking a gradiometer survey, it is important that the instrument is kept in a vertical axis without the bottom sensor brushing against any vegetation. Excessive buffeting results in increased noise levels, and depending upon the nature of the site, may totally preclude good data collection. If in doubt, then consideration should be given to postponing the project until after the harvest; or, the client should be advised to cut and clear the vegetation. Pasture or crops being too tall is probably the greatest cause of fieldwork being cancelled. Long wet grass can cause problems with resistance surveys resulting in poor readings (see the example shown in figure **91**).

Ploughed fields

Deeply ploughed areas and potato fields are particularly difficult to work in. In extreme conditions such fields should be avoided as they will often produce unusual effects, which can easily mask anomalies of archaeological interest. Similarly, tractor ruts can result in spurious anomalies. If the field has been harrowed, recently sown, or the ground rolled, then problems of walking can usually be overcome, but wet, heavy soils will make work extremely difficult and can affect the quality of recorded data. Frozen ploughed soil can be hard to walk over, is difficult to get good electrical contact on and may make GPR survey impossible.

Ridge and furrow

The effects of ridge and furrow are often difficult to predict. However, experience has shown that where there are visible earthworks, it is probable that the gradiometer will not record any major magnetic changes, unless preservation is so good that a topographic effect is produced. In such circumstances, earlier features sealed by the ridge and furrow should still be detectable. By comparison, it has been found that where ridge and furrow has been reduced by ploughing, a striped magnetic effect is likely to result because of the contrasts between the infilled furrows and the former ridges. Obviously this means that the ridge and furrow can be mapped, but as the response is long and linear this information can be stripped away by using a directional filter of some kind (see for example figure **96**).

Base maps and conventional survey

There is little point in carrying out a geophysical survey unless it can be *accurately* located on the ground. This may sound a simple statement but it is fraught with complications for the field surveyor. What is meant by the word *accurate*? For example, what happens when measurements are made from a field post that is 40cm thick; are measurements taken from the edge of the post or the centre; if measurements are not made at the bottom, is the post truly vertical; is the post the same as the corner post depicted on the Ordnance Survey map; how exact is the position on the map? There are no simple answers to questions regarding map accuracy and it is beyond the scope of this book to begin to try to provide solutions. There are plenty of textbooks that deal in more depth with the issues (see Bettess 1992 and Bowden 1999), but a number of matters do need to be covered here, even though they may seem simplistic. It is surprising, if not exasperating, to see geophysical surveys that are tied in by pacing at one extreme and those that depend totally upon geo-referenced points for their relocation at the other extreme. A certain level of common sense and good recording must prevail.

A fundamental starting point must be a suitable base map that is available to all groups or individuals who need to know about the results of the geophysical work. Ideally this will be in digital format so that images and interpretation figures can be placed *in situ* and also printed at any suitable scale. It has to be remembered that a suitable scale for a landscape researcher is very different to that suggested by an excavator. The software that is normally used for the dissemination of this sort of information is a CAD or GIS program as they maintain the integrity of the map at all scales.

The importance of basic surveying within the overall project cannot be emphasised enough. An area survey where the grid has become trapezoidal or where a perfectly set out grid has not been tied in properly reduces the value of the geophysical data. In the EH 'guidelines' the significance of this basic work has been forcefully stated:

> An internally accurate and correctly measured-in grid is crucial to all subsequent survey and, indeed, to the whole project outcome: close attention to this fundamental stage of fieldwork is therefore essential.
> (David 1995)

The vast majority of geophysical survey, no matter which technique is employed, involves the collection of data either along a line, a series of lines or systematically within a predefined grid. These lines or grids need to be placed on the base map so that they can be relocated on the ground, if required, at a later date. In the case of survey grids, these need to be established in the field and retain a rigidity throughout the survey area. There are a variety of ways to carry out this basic survey and the options are discussed below.

It should be noted that in most geophysical surveys for archaeological purposes there is rarely a need to work on very steep slopes. Hence the fact that in small surveys most measurements are made along the ground surface and little effort is made to keep the tape horizontal. If the slope is not too great then there will be no effect in placing the lines or grids on base maps. However, problems can arise on earthwork sites unless a consistent approach is employed.

On larger sites the use of electronic instruments to measure horizontal distances and to establish base points, from which the grids are set out using tapes, also minimises any problems of base plan errors. However, a compromise must be made when a grid is set up over a steeply sloping hillside. If tapes are used then a plan view of the survey results or interpretation will be too large for the area surveyed. This is because each 20m length is measured at an angle from the horizontal and the distance is greater than a horizontal or plan 20m. This will be cumulative downhill and the survey can be seriously incorrect. For the unwary this problem may not be apparent until it is time to place the survey results onto the base map. There is no simple way to amend this problem as the survey will never accurately fit the map. If sufficient tie-in information and a good quality topographic map are available it may be possible to rectify the geophysical information to the base map in much the same way as an aerial photograph can be corrected. For most people the answer would be to establish a horizontal grid where it is known that the grid will fit on an accurate map. However, this is not without practical problems as each 20m grid may be a different size when measured on the ground. In fact the squares may be trapezoidal in shape. How then are equidistant traverses marked within the grid? In such cases common sense must be used to adjust the position of each traverse. As such there is no 'correct' way to undertake survey on slopes but there are many 'wrong' ways. The chances that a survey can be regarded as successful under these conditions are increased significantly if the survey strategy is well documented and the information is passed on to all concerned. In particular this will alert would-be excavators to re-establish the geophysical grid in the same way as it was laid out, therefore reducing possible relocation errors.

Tapes/simple geometry/optical squares

The easiest method of setting out lines and grids is using survey tapes that are available in a variety of lengths up to 100m, though 30m and 50m are the most commonly used. Simple geometry based on Pythagoras' Theorem for right-angled triangles is employed and works well for setting out a few survey grids. This is commonly referred to as the '3-4-5' rule because of the ratios involved. In order to keep errors to a minimum it is usual to measure the hypotenuse of a large triangle. In practice a right-angled triangle is often established with sides of 10m or 20m. The hypotenuses for these two triangles are 14.14m and 28.28m respectively, and the grid can be established using two 30m tapes.

An alternative and very quick way of setting out accurate base lines and right angles is to use an optical square (**colour plate 12**). This is a simple and cheap device in which two prisms are mounted at 90° to each other and set vertically either side of a central window. Consequently there are three 'views' when looking through the optical square; straight on, 90° to the left and 90° to the right. To use the device to set up a right angle a surveyor looks straight through the window at a fixed point, usually a ranging pole, on a base line. A second surveyor then moves a ranging pole into view on one of the prisms set at 90°. When the two poles form a straight vertical line in the two windows, the second surveyor's pole will be at a right angle to the base line. Significant points (10, 20 or 30m grid points, for example) can then be accurately set out between the fixed point on the baseline and the new point. This process can be repeated at the far end of the baseline, or at intervals in between. To minimise errors it is good practice to work within the box that has been set out and lines should not be extended beyond it. Although the method sounds rudimentary, in practice errors as little as *c*.5cm are achievable by expert users over a distance of 100m.

Theodolites/EDM and total stations

For surveys involving areas greater than one hectare, the use of optical or electronic instruments is advised (see David 1995) and the different types of instruments are summarised in Bowden (1999). In archaeological geophysics, because of their speed and relative ease of use, Theodolite/EDM or total station instruments (**colour plate 12**) are most commonly employed. Typically, base points are established at 100m intervals across the survey area and tapes then used to subdivide the blocks into smaller grids that will be used during data capture. The great benefit in using these instruments is that only a few additional measurements are required to tie the grid into a base map. As long as these instruments are treated properly and regularly re-calibrated it is possible to tie into boundaries or structures even at a great distance. This is particularly valuable when the survey is undertaken near the centre of large fields.

Global positioning systems (GPS)

The use of *differential* GPS (**colour plate 12**) allows for the setting out of grids to within 1cm and is arguably the only method that allows truly accurate geo-referencing without regard to potentially inaccurate base maps. The system is discussed more fully in Bowden (1999). As this surveying method is dependent upon the relative position of the measurement point with respect to the known position of the satellites orbiting the Earth, this is clearly independent of local features. As a result, if it is tied-in using

GPS, then relocating the survey in years to come will not be a problem if developments have taken out field boundaries or changed the locality in any way. Crucially, to maintain the advantage of using GPS, then any relocation must use GPS.

The cost of GPS means that their usage is invariably confined to larger projects and is often carried out by a third party, not the archaeological geophysicist. In fact, it is valuable if the base points are set out at the start of archaeological fieldwork to facilitate use by the many groups that have to collect accurate data during an archaeological project.

Whichever method is used, it is important to keep good records of the methods used so that subsequent surveyors know how to re-establish the survey grid. In this respect the use of permanent or semi-permanent markers at the end of baselines in field boundaries is often appropriate as this allows rapid relocation of the grid. However, it is also of benefit to tie into permanent map features, just in case the markers are not as permanent as would be desired.

Strategies for the field

The strategy that is adopted for a geophysical survey is especially dependent upon the geographical scale around which the questions are framed. In an effort to categorise the role of geophysics in archaeology we have suggested elsewhere that the relationship may work on three levels (Gaffney and Gater 1993), reproduced here in figures **38**, **39** and **40**. These levels can be described as 'prospecting', 'assessment' and 'investigation' and relate to an increasing intensity of data collection. This leads to better data control and, hopefully, more secure interpretation. The first two levels are routinely undertaken within the commercial field and are not mutually exclusive; for example scanning an area by itself (Level I, prospecting) is of little value without detailed survey (Level II, assessment). It is perhaps reasonable to assume that the majority of research-type surveys are working at Level III.

In attempting to resolve many archaeological questions the geophysicist will endeavour to survey as large an area as is practical in as great a detail as possible. Time and cost is highly relevant here, especially in commercial work and it is this arena that field strategies have been refined. In particular the use of geophysical techniques in large-scale evaluation projects has evolved considerably during the last decade. In these cases we must search for the archaeology in some other way than by total detailed survey. The basis of all strategies requires a way to establish background levels and gain a good understanding of the effects of the geology and pedology. However, there are many instances where it is not feasible to survey in detail the whole of an area of interest, even though this may be the ideal solution to a problem. Although survey speeds have increased dramatically over the last decade or so, it is still often necessary to adopt a sampling strategy. Fortunately these have been long practised in archaeology and archaeologists are aware of their uses and limitations.

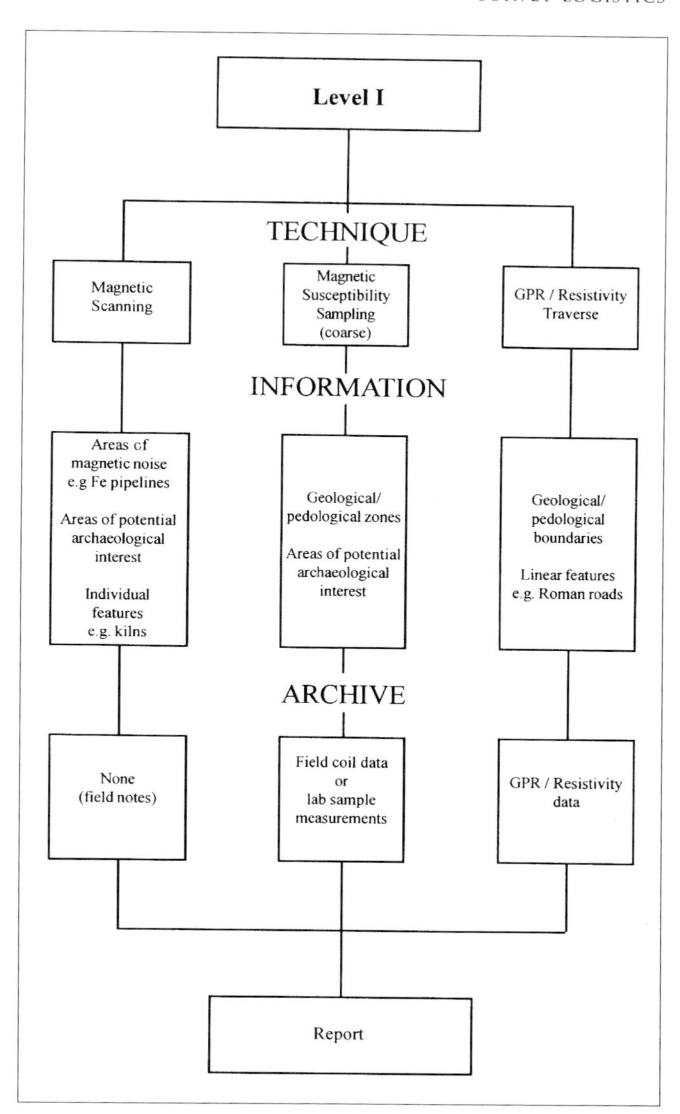

38 *Typical geophysical survey strategies used in Level I – Prospection*

It is apparent that the strategy adopted for researching a small development in a single field will be different from that chosen when investigating a large development affecting many fields. It is likely that the former will go directly to total survey of the area, while the latter will see selection of areas for detailed survey. There may be considerable debate regarding the size at which a sampling strategy should be adopted. Given the time and effort required to set out a survey grid, it is felt that about 2ha is the change point and there are considerable disadvantages in attempting to sample areas of 1ha or less.

If the development is larger than 2ha and not unusually archaeologically sensitive, then some form of sampling is normally undertaken. The approach can

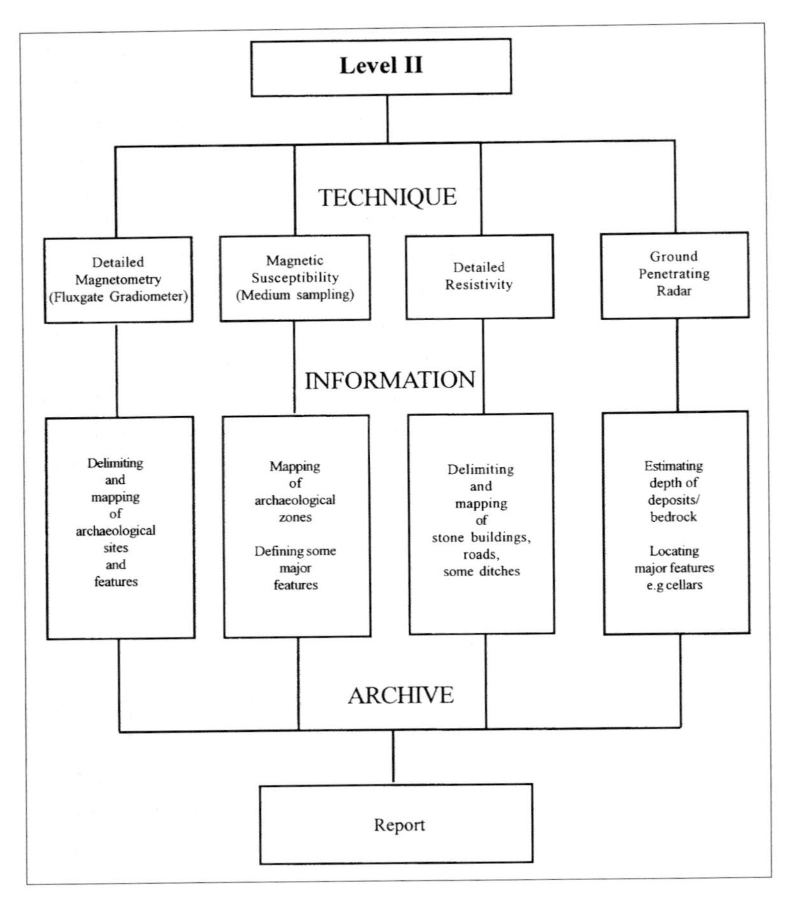

39 *Typical geophysical survey strategies used in Level II – Evaluation*

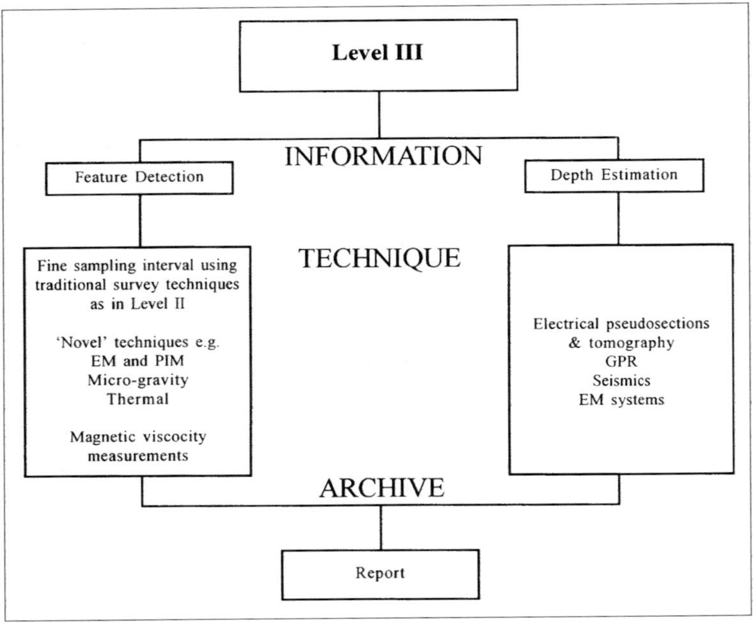

40 *Typical geophysical survey strategies used in Level III – Investigation*

be varied depending upon the archaeology that may be encountered. In Britain the geophysicist will resort to a systematic, random or some form of modified sampling strategy. Examples of sampling schemes can be seen in figure **41**. Of concern in this situation is that the space between the sample blocks should not be too great to miss representative or significant archaeology. Obviously, when you are working in a previously blank area then it is often difficult to judge the correct distance between the samples. This is normally agreed in advance, as is a contingency which must be allowed for to follow any interesting anomalies that run out of the edge of a sample. There are a number of reasons why a sampling scheme may be modified from a purely systematic or random strategy. Firstly, ground conditions may dictate that a sample cannot be placed in the intended position, perhaps a new fenceline has appeared in the field since the map was drawn or ground conditions are patently unsuitable. Secondly, aerial photographic or fieldwalking evidence may indicate zones of interest and not to cover them may leave unanswered questions.

How large should a sample block be?

Sometimes a restricted survey area can lead to interpretation problems. A classic example of such a situation involved the authors during the making of a *Time Team* programme. Our task was to try to locate a Roman road and associated features on a level field at Ribchester, Lancashire, within a known Roman site. After surveying three 20m squares the initial results (low resistance linear anomalies) indicated what

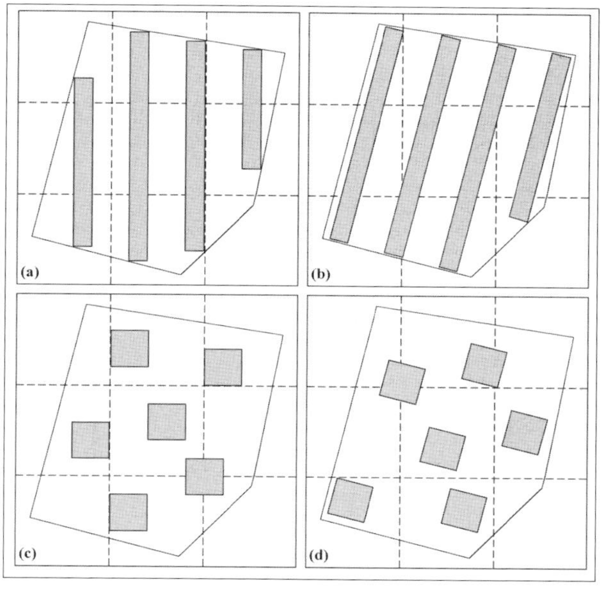

41 *Examples of strategies that can be used to sample areas of possible archaeological interest using detailed survey. The decision to use either transects or blocks are normally based upon the nature of the potential archaeology. Using the National Grid as a baseline (a and c) may cause difficulties in irregular shaped fields. Grids aligned to field boundaries (b and d) may be preferable*

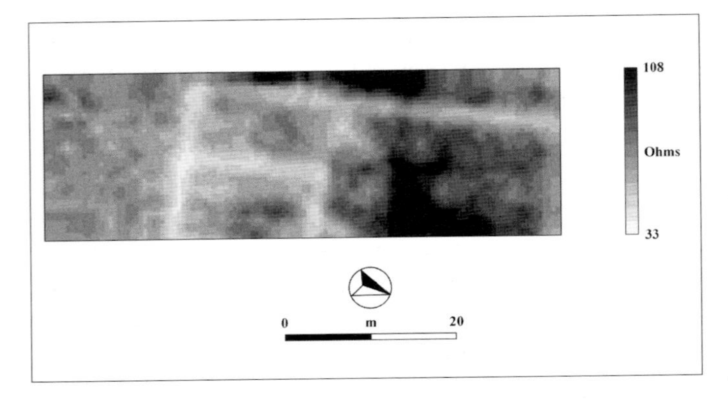

42 *Twin-Probe resistance survey, 1 x1m, within the Roman site at Ribchester, Lancashire. It was assumed that the low resistance anomalies (white) were possibly ditches . . .*

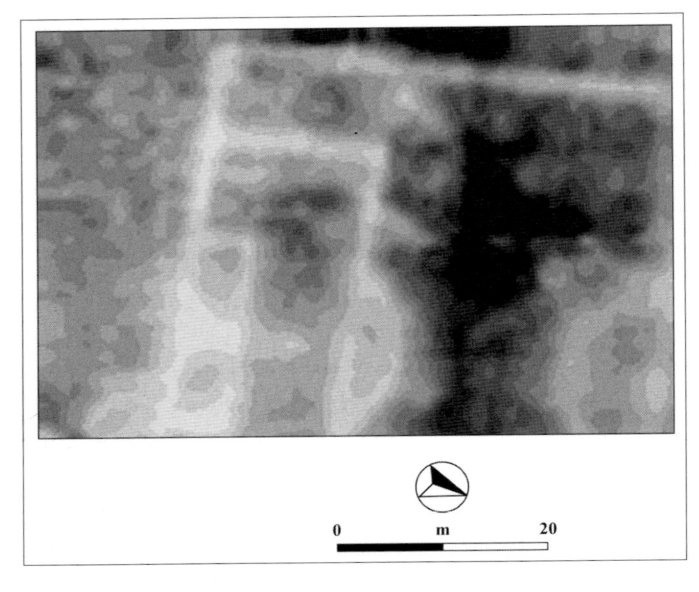

43 *. . . when the area was surveyed it became clear that the 'ditches' were in fact the markings of a former soccer pitch. While the pitch is not believed to be Roman in date, the high resistance anomaly (black) cutting across the playing field is likely to mark the position of a former Roman road*

appeared to be straight ditches that turned through right angles. Our instant 'on camera' interpretation suggested these were Roman ditches indicative of enclosures perhaps appended to the road – which still remained to be found. The results at this stage are reproduced in figure **42**. The survey area was subsequently expanded by another three grids and to our surprise some of the straight 'ditches' had curves attached to them. Also two large arcs became apparent, along with smaller recti-linear features. When a single low resistance response was noted between the two arcs our interpretation became rapidly, and radically, different (**43**). The resistance survey had produced a clear plan of a former soccer pitch complete with the penalty spot! The former white lines (no longer visible) had been repeatedly marked with lime and this had affected the moisture-retaining capability of the soil in a metre-wide strip. On reflection, the results do show a probable road crossing the survey area but the data serves as a good example of drawing incorrect conclusions before a large enough area has been surveyed to establish the background.

There are, however, other ways to establish where to position sample blocks. These involve some form of rapid assessment over the fields in question. Rapid assessments are normally undertaken by 'scanning' with a fluxgate gradiometer or a widely gridded survey using a magnetic susceptibility field coil. Both techniques have their advocates and it is certain that there are advantages and limitations to each approach.

Probably the most frequently used approach is scanning with a fluxgate gradiometer. When undertaking a scan an *experienced* operator endeavors to assess the background level of magnetic response as he or she walks along traverses, usually spaced 10-15m apart. In large surveys it is often best to undertake the work using two or three people operating in conjunction. This helps alleviate boredom and immediate feedback is available from adjacent traverses. An experienced operator will normally be able to identify anomalies as weak as 2nT. This 'scanning threshold' will vary depending upon the background noise; this may be due to natural soil variation/geology, or as a result of human agencies in the shape of manuring or modern rubbish spreads. When an anomaly is found, a marker, usually a cane or a peg, is pushed into the ground at that point. The position of the anomaly is also marked on a map so that an adequate strategy for investigating these points can be worked out. The strategy always involves detailed survey using the fluxgate gradiometer in sample blocks over the targets. In some cases, perhaps if few targets are found or if the spatial sampling is poor, apparently 'blank' areas will also be sampled by detailed survey using the same instrument. Occasionally a variation on scanning has been tried where data is captured along long lines. This 'Recorded Line Scanning' is not very efficient as the operator cannot easily stop within a traverse to analyse readings of interest.

From the earlier chapters it will be apparent that some features will not produce anomalies as great as 2nT. Sampling 'blank' areas will act as a check to see if significant anomalies are present that are lying below the scanning threshold. It should be stressed that 'scanning out' anomalies can be very difficult and, on sites that have had a low level of activity during their archaeological life or on sites where there is a significant igneous presence, even experienced personnel may have limited success in detecting *areas* of interest, let alone individual anomalies. It is easy to be confused when there are beds of magnetic gravels in the search area, or the magnetic contrast between fills and subsoil is too weak to produce detectable scanning anomalies, as in deep undifferentiated sands. To get a feel for the likely strength of anomalies in a new area, it is sometimes helpful to collect some detailed survey over known buried archaeology. Normally, this would be over cropmark evidence or similar good quality information. However, if this does not exist then it is impossible to do an assessment in the field. In order to work out the chances of success then we have to resort to experience with similar geologies.

As an alternative to fluxgate gradiometer scanning, magnetic susceptibility can be used to identify zones of high archaeological potential. This method of assessment is very different from using a fluxgate gradiometer, in that data are collected over a grid and have to be interrogated prior to selection of areas for survey. In magnetic susceptibility sampling, measurements are recorded systematically across an area; data are usually collected on a 5-20m sample grid and a map of topsoil susceptibility is produced. The data can either be logged directly into a hand-held computer or written down and typed into suitable software. By displaying the data around a statistical variation a number of areas of increased susceptibility or 'hotspots' will be apparent. These areas, and 'blank' control areas within the data set, can then be assessed with the fluxgate gradiometer using sample blocks. The strength of this approach is the fact that data can be analysed later as opposed to the scanning approach that only notes potential targets in the field. It is important to explain why the sample blocks of detailed survey should be placed in particular areas, and it is obviously easier with an image of real data rather than a map showing scanned anomalies which is in effect an interpretation. However, there are also a number of disadvantages with this approach. In particular it is known that the history of land use and modern use will produce spurious variations, as will the variation in many natural/pedological factors. In terms of interpretation, although there are laboratory methods to discriminate 'natural' from 'anthropogenic' soils, volume susceptibility does little more than chart variations in topsoil susceptibility. The worst case could be that an increase in susceptibility may simply represent totally ploughed out remains and in this scenario detailed fluxgate survey would reveal nothing. As with the fluxgate there are similar problems in locating ephemeral sites (i.e. 'ritual') and, given the large sample interval, discrete small sites. Two other considerations that should be taken into account are the fact that susceptibility survey takes longer than scanning with a gradiometer, but can be undertaken by operators with little technical ability.

Both scanning techniques produce good responses over large Iron Age/Romano-British settlements as these are usually reasonably large in size and the various strata often exhibit well enhanced and spatially variable magnetic susceptibility. The results are more uncertain when assessing small or short-lived sites, especially when settlement or industrial activity is absent. Irrespective of which method is used for rapid assessment it is always preferable to investigate large sample areas during the follow up detailed gradiometer survey. This allows a greater understanding of background variations that may be the result of geological and topographical factors. As a broad guideline, a 60 x 60m area is preferred as a sample block, and areas less than 40 x 40m in extent are often uninformative. If transects (i.e. sample strips) are used to sample an area, it is suggested that they should be a minimum of 20m wide. If the transects reduce below this size then the information they produce is very low grade and often the interpretation is no better than indicating the presence or absence of anomalies of interest.

Sample and traverse intervals

Up to this point questions regarding the choice of sample intervals along each traverse and the distance between traverses have been largely ignored. Historically, choices were often limited due to the lack, or limited capability, of data capture. The use of 1 x 1m collection strategies used in early work had their basis in earlier 3 x 3ft surveys. In fact digital data sets with smaller distances between samples did not become achievable until data logging was commercially available in the 1980s. Despite the logging facilities associated with modern instruments, resistance data collection is still acceptable in most cases at 1m spacing. This is largely due to the slow speed and the fatigue associated with manual insertion of probes. Although the multiplexer arrangement used by the Geoscan Research machines can achieve notable increased data capture, for the sample interval to be routinely 0.5m or less then a robust wheeled system must be devised. Until that is realised then 1 x 1m surveys will remain the norm, while 0.5 x 0.5m surveys will be conducted on small areas where great detail can be expected, such as features within historic gardens. The strategies for collecting magnetic data have seen large changes over the last decade. Although magnetic surveys were also traditionally recorded at 1 x 1m intervals, the fact that it is effectively a continuous measuring device means the interval between samples can be easily reduced. Until recently the high cost of data logger memory, the slow downloading times between instruments and computers, and the intensive use of the computer resources when processing large data sets, meant that two samples per metre along each traverse was regarded as normal in most cases. Given the antiquity of many of the surveys reproduced in this book, the majority of the data sets were collected at two samples per metre for gradiometry and one sample for resistance data. However, these hurdles have largely been overcome and increasingly magnetic survey data collected at fewer than four samples per metre is unusual. The benefit of collecting increased numbers of data samples is the ability to define smaller features and find out more about the background noise that is due to the soil and/or instrument. The latter will allow greater and more powerful processing options. But, the key to sharper images is not just increasing the number of measurements in one direction. Decreasing the traverse interval to 0.5m is very effective for producing high quality maps and also allows significant improvement in automated interpretation programs. Systems such as the new dual fluxgate configuration and the wheeled caesium vapour instrument, described in chapter 3, are the key to increasing the number of traverses and it is likely that developments such as these will drive the collection of data sets with close sample and traverse intervals. At present the standard of 1m traverse intervals is likely to remain adequate for the majority of surveys, although research (Level 3) surveys will increasingly collect data at closer intervals.

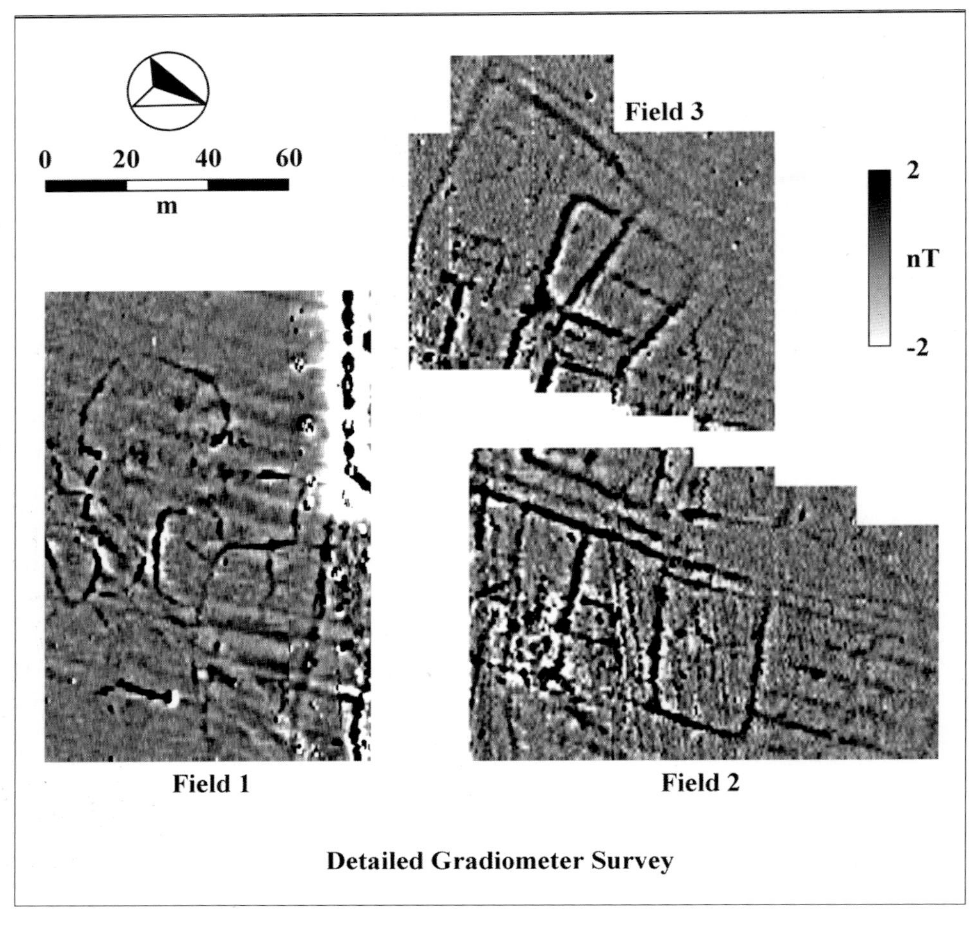

44 *Fluxgate gradiometer data (1 x 0.5m) collected at Normanton, Lincolnshire. The results indicate a number of phases of activity. An increase in magnetic noise at the southern edge of Field 2, which was interpreted as a building, was subsequently confirmed by excavation*

Rapid assessment: two case studies

It is very difficult to assess the true effectiveness of search techniques as very few areas that have been subjected to this form of analysis are then excavated in their entirety. Archaeologists usually use the evidence from the geophysical report to guide the placement of trenches. Other evidence such as surface collection, place name, documentary and cropmark analysis will also be taken into consideration during trenching, but geophysical evidence figures highly in deciding exactly where the trenches should be placed. This in part is due to a presumption that the geophysical techniques have located important elements of the buried archaeological landscape. Furthermore, it is believed that the anomalies found during the fluxgate scan or magnetic susceptibility survey will be representative of the archaeology.

1 (Above left) *Photograph of the Twin-Probe array in action. The instrument sits on the frame which is attached to the 'mobile' pair of probes, one current and one potential. The 'remote' pair of probes are positioned at 30 times the distance between the 'mobile' probes*

2 (Above) *A Geonics EM38 in action*

3 (Left) *Seismics. The red cable connects to a line of geophones, the first of which is just visible alongside the red peg*

4 (Above left) *A Geoscan RM15 mounted on the frame with an MPX15 multiplexer unit attached below*

5 (Left) *A Geoscan FM256 fluxgate gradiometer which incorporates the old ST1 sample trigger inside the main box of electronics. The trigger is now operated by means of the 'Start-Stop' button. Batteries are now housed below the instrument*

6 (Above right) *A dual gradiometer system. One instrument acts as a 'Master' and controls the speed of data collection, but both instruments record data that have to be separately downloaded*

7 (Above) *A Bartington Grad 601-2 system. The lightweight instrument is counterbalanced using a harness system*

8 (Left) *A Scintrex Smartmag SM-4G set up as a 0.5m gradiometer.* Photograph supplied by English Heritage

9 (Below) *A caesium vapour magnetometer. The sensors are mounted on a cart which is pushed across the field.* Photograph supplied by VIAS Geophysical Prospection, University of Vienna

10 (Left) *A Bartington magnetic susceptibility field coil. The instrument is 'zeroed' between station intervals by inverting the coil in the air. Measurements are taken by placing the coil on the ground surface*

11 (Below) *A Sensors and Software Pulse Ekko 1000 GPR system linked to an odometer which keeps track of the distance surveyed*

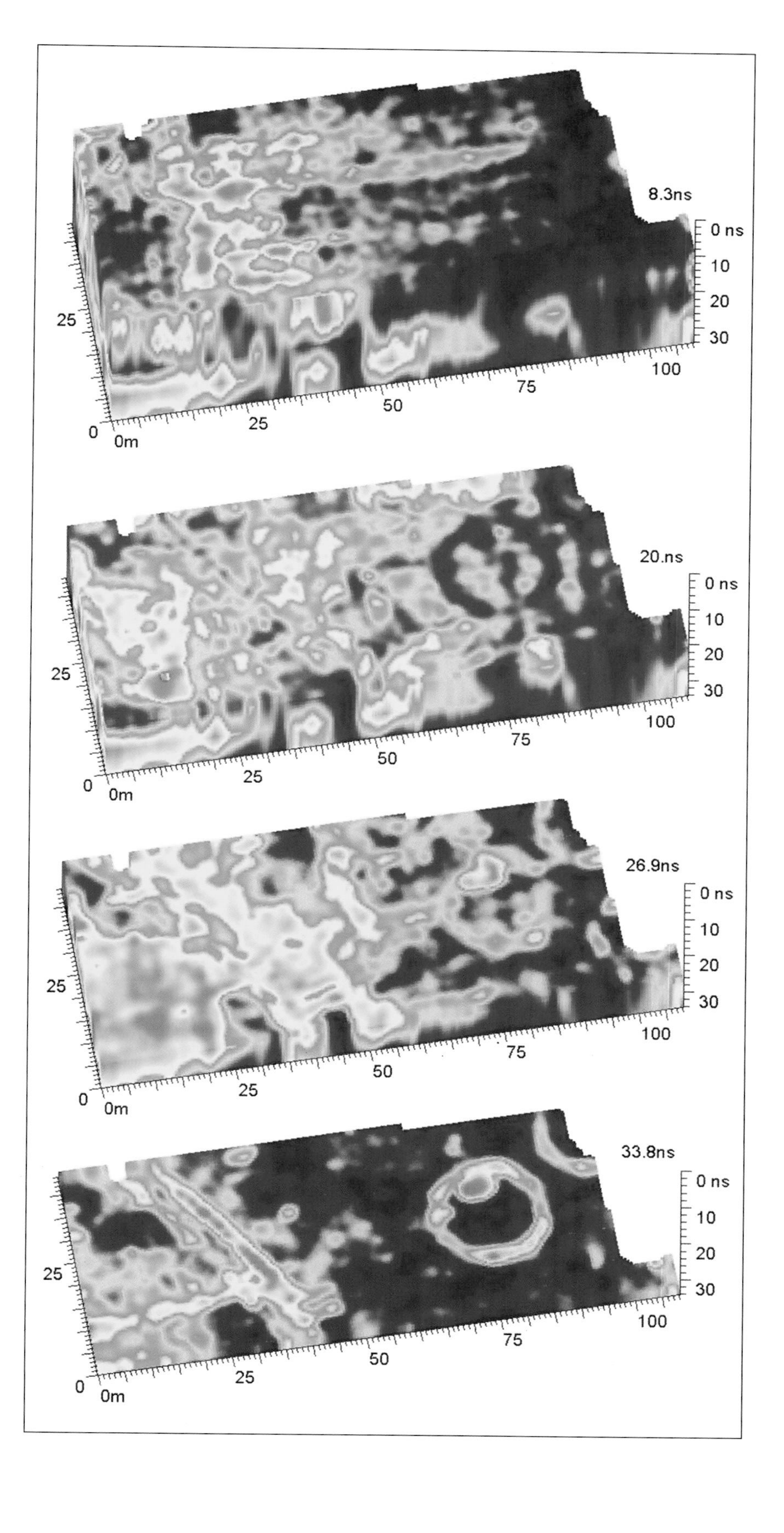

8.3ns
0 ns
10
20
30
25
0
0m
25
50
75
100
20.ns
0 ns
10
20
30
25
0
0m
25
50
75
100
26.9ns
0 ns
10
20
30
25
0
0m
25
50
75
100
33.8ns
0 ns
10
20
30
25
0
0m
25
50
75
100

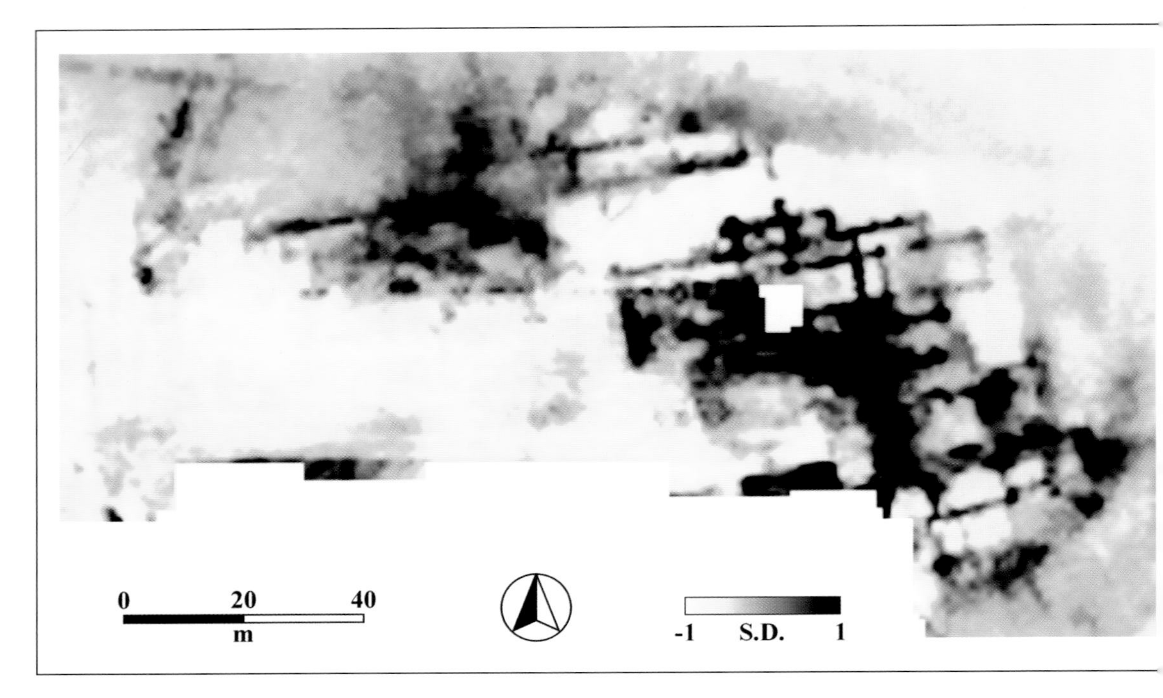

23 *Athelney Abbey, Somerset. Twin-Probe resistance data. 1 x 1m*

24 *In this image Juerg Leckebusch has produced a 3D perspective view with shaded relief. The majority of the GPR data have been excluded to allow a visualisation that is also an interpretation. It shows construction details in the upper and lower parts of a medieval wall (red), which is cut into a surface (blue) within the Prediger Church in Zurich. Undulations in the buried surface are believed to be due to small boulders*

It is unlikely, however, that geophysical techniques will discover every site or indeed every feature within a site.

As there is some debate as to the effectiveness of the two rapid techniques (scanning and magnetic susceptibility sampling) it is worth looking at the information that they can produce. In an attempt to understand this information better, some results from a site at Normanton (England) are of interest. Essentially the whole of the area shown in figure **44** was surveyed in detail using a fluxgate gradiometer. The data show a plethora of anomalies that indicate a large site and it can be argued that the gradiometer data suggest that there are two distinct phases at the site, with a large area of regular enclosures in Fields 2 and 3, while Field 1 has a significantly different pattern of ditched features. The former are Romano-British in appearance, while the latter are reminiscent of Iron Age remains. An area of noise derived from structural deposits has been identified in Field 2 and excavation immediately after the survey revealed walls of Romano-British date. On the basis of these results it was suggested that we may be seeing an interface between

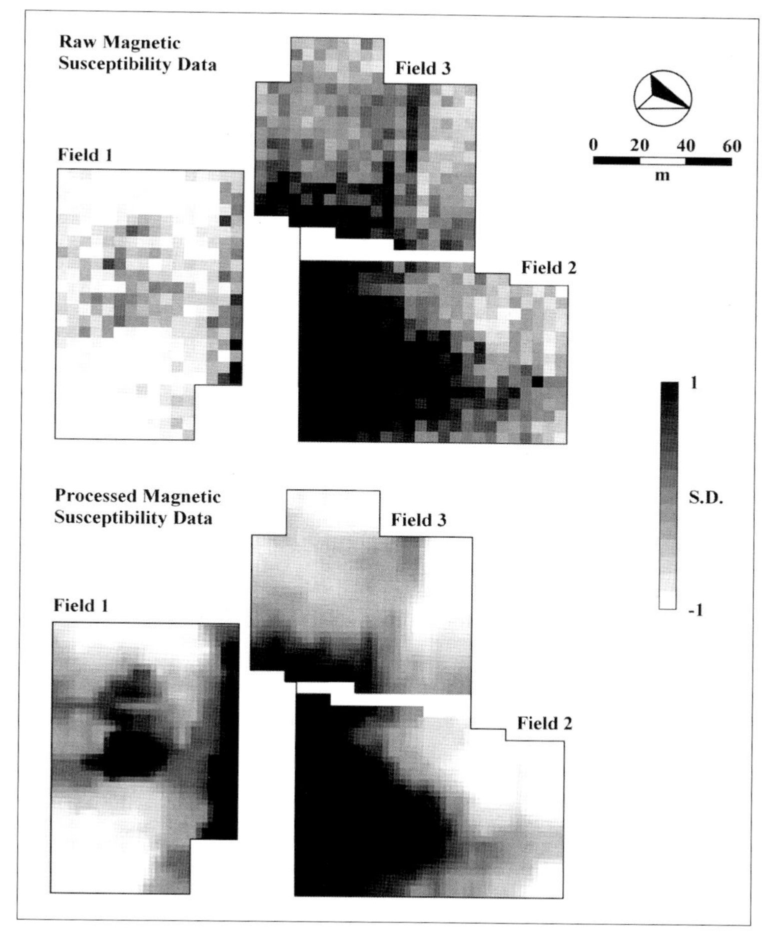

45 *Raw and processed magnetic susceptibility data from Normanton. The data were collected on a 5 x 5m grid using a Bartington coil system with a MS2D sensor. Compare with* ***44***

Iron Age and Roman occupation. While a number of skeletons were discovered during the excavations no evidence for the burials can be found within the gradiometer data. Other elements to note include the ridge and furrow and modern ploughing overlying the archaeology, the pipe that dominates part of the data set, and the variation in the strength of the magnetic response across the site.

A magnetic susceptibility survey (MS) was undertaken over the three survey areas. The measurements, at 5m intervals, were taken with a field coil and it was hoped that the survey would aid the definition of site size as well as giving information on internal (or intra) site zonation.

The first interesting factor to note is that the topsoil susceptibility values collected in Field 1 are consistently lower in value than the two other fields, reflecting the presence of grass cover as opposed to bare, ploughed earth in the other fields. This is a typical non-archaeological variation and must be corrected in some way. For this study the data have been visually 'corrected' by applying a multiplication factor to Field 1 and the whole of the data set has been de-spiked and interpolated. This gives the data set a smoother appearance (**45**).

It is clear that the susceptibility corresponds very closely to the core of the settlement in Field 2. The analysis of the other two fields is more ambiguous. In Field 1 the MS data neatly emphasise the zones of strong sub-circular anomalies in the southern part of the survey area. This suggests that they are inherently different from the main block and points towards a different use. Although we would argue that, given the pattern of the gradiometer response, this is likely to be the result of a chronological division as well as a change in use. In Field 3 the results are again revealing as the high susceptibility is confined to the eastern boundary that divides Fields 2 and 3. This presumably supports the conclusion from the gradiometer data that the enclosures in the western part of the survey reflect agricultural rather than settlement use.

The value of the MS data in establishing the site limits is more debatable as there are areas of significant archaeological anomalies that do not produce an enhanced MS signal. This is most noticeable in Fields 1 and 3. This fact is particularly important if MS is used to locate sites as significant elements, such as the western enclosures in Field 3, which would not have been located if a narrow transect was placed through this area. Of considerable methodological significance is figure **46** which shows the gradiometer anomalies in excess of 2nT as well as the MS data. The scanned fluxgate anomalies shown are all values above the scanning threshold, i.e. all the responses that could be found by scanning. The black lines are 10m scanning traverses and where they intersect with values above 2nT it can be assumed that these anomalies could be located. A cursory analysis of this data clearly indicates that all the significant elements would be found by scanning, including some that were not well indicated on the MS survey.

The MS data has located the main site in a convincing manner, as well as supporting the interpretation of two phases within the gradiometer data. The MS data do not indicate the presence of some of the peripheral, but still important, elements of the site. It has been shown that these elements would have been found using scanning. However the MS is of value in the validation of activity zones suggested by the detailed gradiometer data.

One more aspect that should be mentioned is how much detailed survey should be undertaken after a rapid assessment has been done. If the project is

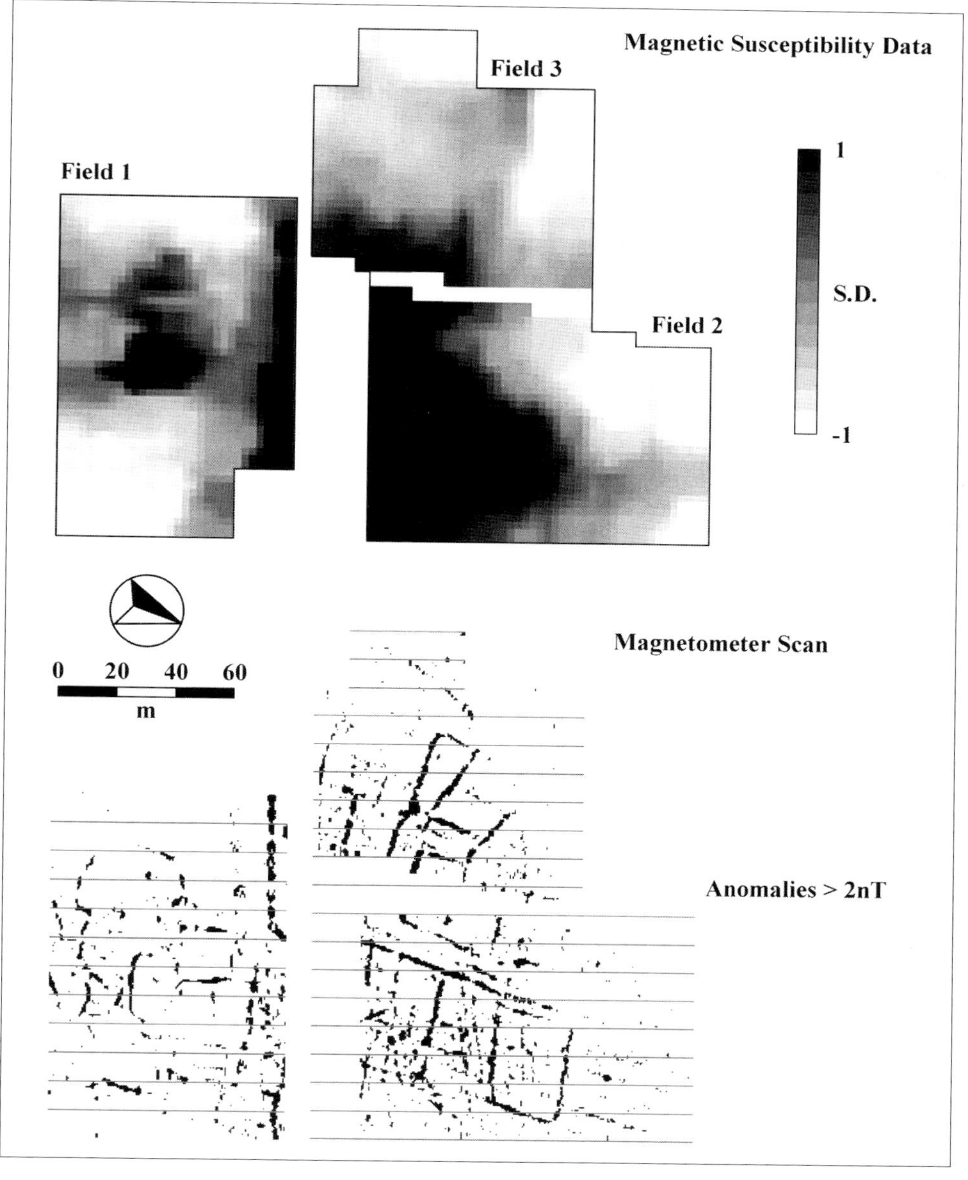

46 *Two strategies for rapid assessment. The fluxgate scanning lines overlie a plan of all values greater than 2nT within the detailed survey data. It can be seen that the magnetic scan would locate all major elements of the archaeology found during the area survey*

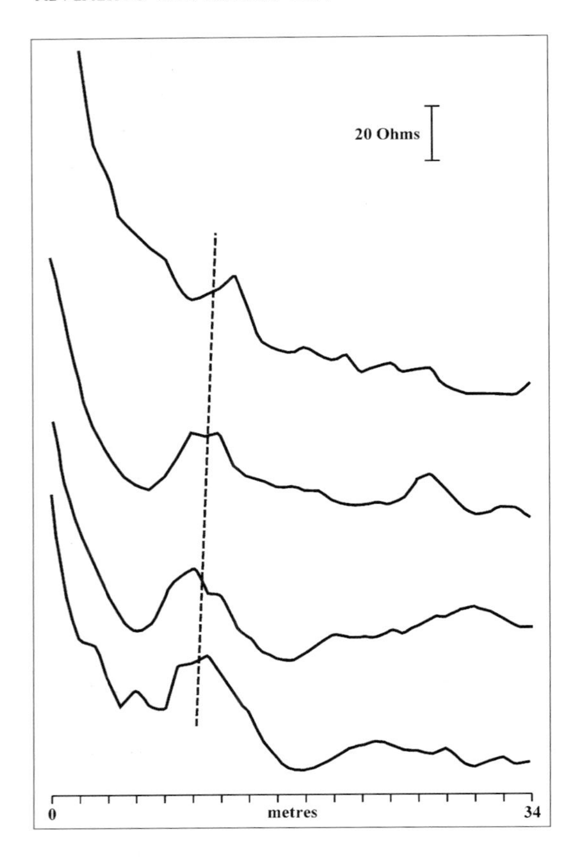

47 *An example of Twin-Probe resistance survey transects used in a 'scanning' exercise. The lines are 10m apart and the course of a Roman road is shown by the dashed line. The signal from the road overlies a resistance trend due to either geological/topsoil variation. Natural variation, as seen in this example, makes scanning with resistance very difficult.* Data courtesy of Roger Walker

research oriented then this question is largely irrelevant as the scan will be used to 'cherry-pick' the positions of high potential. If it is undertaken to evaluate a piece of land prior to possible development then a more structured approach must be followed. Essentially this is usually dependent upon a number of factors; these include how much archaeological potential the area has, the nature of the intended development of the land and what has been specified as a minimum by the local archaeologist. The last point is highly variable as the emphasis placed on archaeological geophysics can vary around the country. As a rule of thumb, we normally expect detailed survey to cover 10-50 per cent of the total area analysed by rapid assessment. Most briefs/specifications also allow for an additional contingency should extensive archaeological anomalies be found. The most extreme case would entail total coverage with detailed surveying after a rapid assessment has taken place. A fluxgate gradiometer scan or magnetic susceptibility field coil survey without a detailed survey to test the results is of little value.

While it has been suggested earlier that it is very difficult to scan with a resistance meter, it is possible when features of a known, or suspected, orientation are to be traced. The resistance data shown in figure **47** illustrate search transects to find and trace a suspected Roman road. The recorded line scans are 10m apart and show a broad high resistance anomaly set within a broader trend. The former is the road and the latter is probably a geological or pedological change. The varying background exemplifies the problem of scanning with the resistance technique, but also shows the use of recorded transects. As a result of the scan it is now known where the road is and area survey could be undertaken to reveal the detail of the road and the archaeological context that surrounds it. It is also possible to undertake a series of scans in a similar manner using electrical imaging or GPR.

Summary of strategies for large areas of land

100 per cent detailed survey. Complete coverage of the whole area. Although this is desirable it may not always be achievable due to the allocation of resources.

Systematic sampling using blocks or transects. An initial percentage of the land would be sampled and an agreed contingency would be used if potential archaeology was found.

Random (or semi-random) sampling. Again an initial percentage would be sampled, normally using randomly positioned grids and an agreed contingency would be used if potential archaeology was found. Although random, the blocks would still use a common grid to allow seamless joining of adjacent areas using the contingency.

Rapid assessment by fluxgate gradiometry. The whole area assessed by scanning in which experienced operators walk along parallel transects separated by 10-15m. Once the anomalous areas have been found then an initial percentage would be sampled using blocks positioned over areas of interest and 'blank' areas could also be surveyed. These are on a common grid so that the sample blocks can be joined if needed. An agreed contingency is often, but not always, in place.

Rapid assessment by volume magnetic susceptibility and magnetometry. Assessment of the entire area is carried out by taking measurements in a systematic fashion. This normally involves taking measurements at 5-20m intervals. The data are analysed and an initial percentage of the whole area would be surveyed with blocks positioned over zones of enhanced susceptibility, as well as other areas. A contingency is allowed for if required.

5

POST-SURVEY ANALYSIS

DATA AND IMAGE PROCESSING, DATA DISPLAY, INTERPRETATION AND REPORT WRITING

Data and image processing

As a general rule, the best policy in all surveys is to collect good quality data in the field, data that are free of, or have minimal, operator or instrument error. Given the unpredictable size and orientation of archaeological remains, the aim should be to keep data processing to an absolute minimum. Naturally, each geophysical technique employed will require specific routines to be followed, but the person doing the processing should never lose sight of the original data. The old adage 'if it isn't in the raw data it isn't there' should be borne in mind. The mathematical manipulation of the readings can easily remove anomalies of potential interest or, equally significantly, create spurious responses that may be interpreted as archaeologically significant (Aspinall 1992). This warning is constantly repeated by experienced operators who are only too well aware of the mistakes that can be made:

> As with all computer modification techniques, alterations of the data should only be applied for specific reasons and may not be warranted for all data sets. One should never attempt to use processing programs 'off the shelf' without understanding the implications of each data manipulation technique.
>
> (Conyers and Goodman 1997 p.78)

The primary purpose of data processing, therefore, should be to assemble the data collected in the field into a coherent form and to carry out mathematical processes on the readings, so that the results can be displayed in a meaningful way prior to interpretation being carried out. A secondary aim should be to remove any known errors in the data. In order to carry out any significant processing the data need to be in a digital format.

At this stage it is worth noting that there is a difference between *data processing* and *image processing*. The former involves manipulation of the actual data through a series of steps that vary according to the type of data being analysed. Data processing can specifically help to highlight additional information that is hidden within. By contrast, image processing, in simplistic terms, involves the manipulation of data to produce what might be termed a visually pleasing image of the results. To take an example: in a resistance survey a high pass filter can be used to remove broad geological changes from a dataset and thus highlight variations that are associated with archaeological features; in this case (data processing) a new dataset is created. An alternative approach is to take a greyscale image of the resistance data and in effect shine a light across the data to highlight the archaeological features. In this instance (image processing) the same dataset is maintained throughout the process. It is fair to say that, as processing routines become all encompassing, the boundaries between data and image processing become increasingly blurred (**48**).

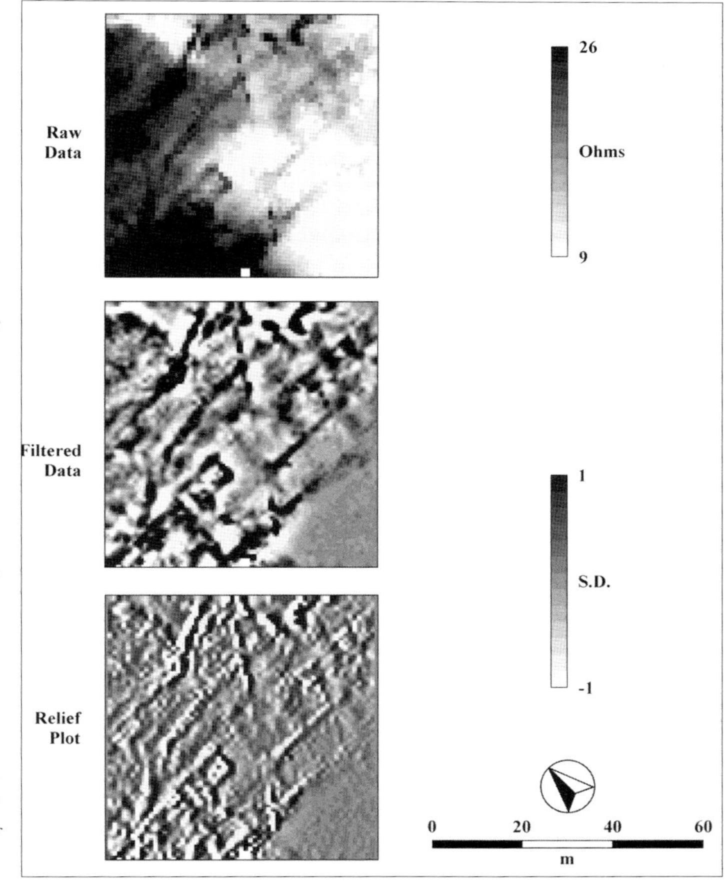

48 *Twin-Probe resistance data from Turkdean Roman villa, 1 x 1m data. The raw resistance data shows a number of linear anomalies that are obscured by the varying background response. To remove this background the data have been filtered in the middle image and displayed as a relief plot in the lower image. The former results in a new data set and is an example of data processing, while the latter is a display of the original data and is an example of image processing*

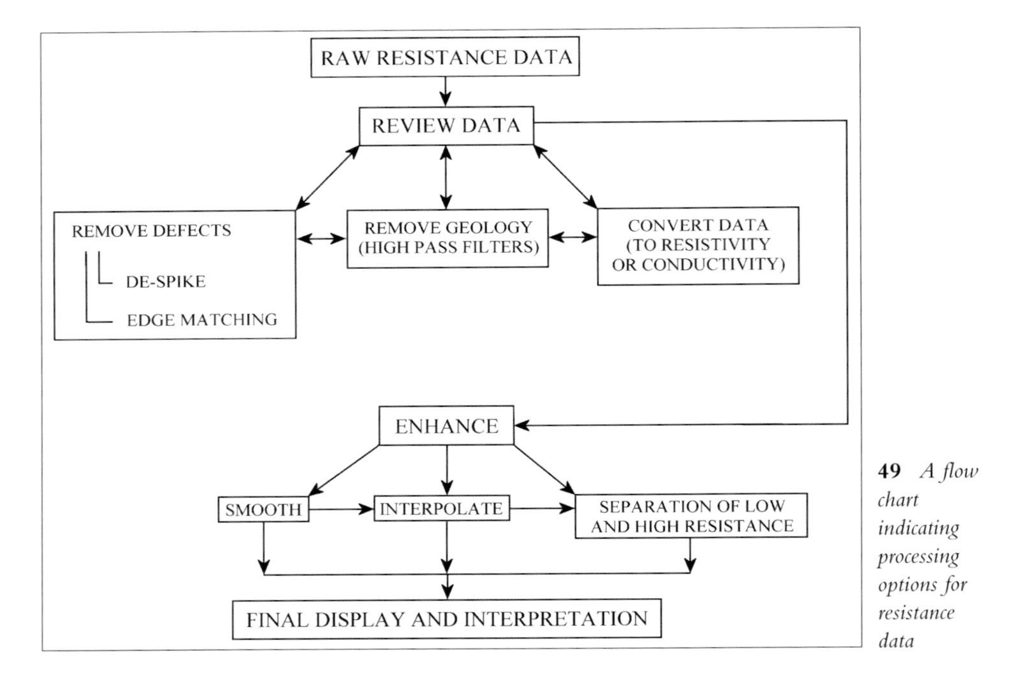

49 *A flow chart indicating processing options for resistance data*

In their Geoplot software manual Geoscan Research produce very clear guidelines for processing resistance, magnetic, magnetic susceptibility and electromagnetic data. Graphical summaries of the processing steps suggested in that document for resistance and magnetic data can be seen in figures **49** and **50**. Even in ideal survey conditions it is extremely rare, after reviewing the data, to proceed immediately to a final display of the data. It is worth noting that there is a certain amount of trial and error with the sequence of processing; there can be no hard and fast rules. However, some steps are usually required at an early stage; for example it is normal to remove poor data points otherwise they will skew the summary statistics as well as create false anomalies. Other processes, such as interpolation, tend to be optional. Interpolation cosmetically enhances the data, giving it a smoother look; although this enhancement increases the number of points to display it does not increase the number of real data points. Interpolation leads to significant increases in processing and display times and when it is used it is usually the last stage prior to final display.

Conductivity data are treated in a similar fashion to resistance data, with the exception that, due to the way the data are collected, some inherent errors are similar to magnetic data and are dealt with accordingly. For example the instruments are prone to drift, so a correction may have to be applied. EM magnetic response and magnetic susceptibility data are processed largely as magnetic data. However, care should be taken when stripping out background variations as this can remove potentially significant archaeological information.

With regard to processing GPR data the most authoritative source of information is provided by Conyers and Goodman (1997), although instrument

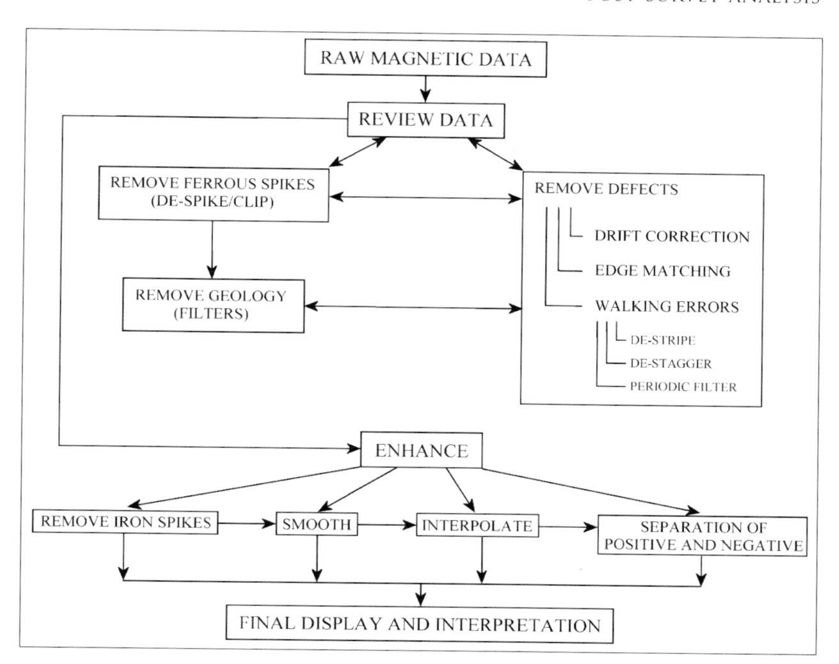

50 *A flow chart indicating processing options for magnetic data*

manufacturers also provide excellent information, e.g. Annan (1997). On collection the data are either in a true raw format or have already been subjected to filtering in order to facilitate real-time viewing. Whatever the case, a number of processes are necessary to edit the data prior to further data processing. These include reorganising, merging, repositioning and renaming the data files, and, where appropriate, the inclusion of topographic information. Subsequent processing is designed to remove noise or interference, and correct the horizontal and vertical scales of the raw data. Various routines are followed including temporal and spatial filters, background subtraction, time gain adjustments and migration. The sequence of the processing steps is open to debate, but the most important fact is that the data should be as 'clean' as possible prior to the production of time slices. If they are not, then many of the anomalies seen in the slices will be the result of imperfections in the data.

Software packages

There are a variety of software packages available, each with their own strengths and weaknesses, although the final output is especially dependent upon the operator's knowledge and experience. It is not our intention to describe all the packages in detail as they are being constantly modified. We will simply highlight some of the features of the main programs used to correct and display data from the more commonly used techniques, namely, magnetometry, resistivity and GPR. For those interested in learning more it is recommended that they visit the websites of the 'main players'.

The packages that are used for post survey work tend to fall into five categories:

Data processing
Data display
Image processing
Interpretation/drawing packages
Desktop publishing

Some concentrate on one category while others are more comprehensive. For instance, the Bartington software that comes with the Grad 601-2 instrument is simply designed to download data from the instrument into a computer. It attempts no more; the philosophy is that there are many other packages available to carry out the next steps of processing. By contrast, the 'Geoplot' package from Geoscan Research is a complex piece of software that not only downloads the data, but also has options for almost every conceivable data processing operation required and some image processing tools. In addition, the package has an interpretation/drawing facility cum-simplified desktop publishing component for producing finished diagrams.

Resistance, magnetic and EM including magnetic susceptibility software packages written specifically for archaeological data processing are available from a number of commercial sources, though probably the most commonly used software comes from the main manufacturer of these instruments, Geoscan Research, mentioned above.

Most of the commercial GPR software tends to be developed from seismic refraction packages used in exploration geophysics. All GPR manufacturers, such as GSSI, Mala, and Sensors and Software have powerful processing packages. Although they share many fundamental processing aims, they differ in their ease of use. Undoubtedly the latest version of Radan (v.5) from GSSI provides an easy route from data collection to animated time slices. The software can calculate velocity by analysing hyperbolic responses at differing depths and then apply a variable correction through the data. Apart from the instrument manufacturers, a number of software specialists also produce good GPR packages. One of these is by Dean Goodman who has published many articles on visualisation for archaeological purposes (www.gpr-survey.com). It is also possible to try out GPR processing on data from the archaeological site of Petra, both data and software can be downloaded via an excellent online publication at e-tiquity.saa.org/%7Eetiquity/title1.htm (Conyers *et al.* 2002).

Other commonly used packages that will readily accept and process data include Surfer and Geosoft. These are primarily all-round performers that have very powerful algorithms for mathematical processing of data, and numerous display formats. They are not strictly geophysical packages as they do not contain algorithms that fix the many small errors that occur during data collection, such as grid or line shifts. Some practitioners download and correct the data using the manufacturer's software and export the data to either Surfer or Geosoft for display.

Data display

When dealing with large datasets it is important to be able to visualise the detail within the complexity that is often present (**51**). When visualising data there can rarely be a single 'best' display option, and given a digital data set the number of ways that a data set can be viewed is enormous. A variety of graphical options are commonly used including:

Dot density

This was one of the earliest and most popular options used for displaying magnetic and resistance data (Scollar 1965); the main advantage is the ability to convert data into a plan form. The display also benefits from providing a limited amount of information on the strength of the anomaly, which helps with interpretation of the data. As the name implies, dots, or pixels, are used to represent the numerical values between selected minimum and maximum cut-off levels. Each reading is allocated a unique area dependent on its position on the survey grid, within which numbers of dots are randomly placed. Normally any value that is below the minimum will appear white, whilst any value above the maximum will be black – the reverse option is also available. Values that lie between these two cut-off levels are depicted with a specified number of dots, which increases in proportion to the value of the reading. The main limitation of this display method is that multiple plots have to be produced in order to view the whole range of the data. It is also difficult to gauge the true strength of any anomaly without looking at the raw data values. However, skilled data operators can use this display to simplify the data into a form of interpretation.

XY traces or stacked profiles

These involve a line representation of the readings and were the most common representation of magnetic data from the late 1960s to the 1980s, when chart recording systems were employed. Each successive row of data is equally incremented in the Y-axis, to produce a stacked profile effect. There is usually an option to incorporate a hidden-line removal algorithm, in order to blank out lines behind the major peaks. The advantages of this type of display are that it allows the full range of the data to be viewed and shows the shape of the individual anomalies. The display may also be changed by altering the horizontal viewing angle and the angle above the plane. By joining the values in the X-axis, as well as the Y-axis, a 3D wire mesh can be created, which itself can be viewed from differing angles. It is possible to fit a surface to this mesh in order to produce a plot similar to a digital terrain model (Bowden 1999). The output may be either colour or black and white.

Greyscale/half tones

Since the arrival of more powerful computers to carry out data processing (see for example Aspinall and Haigh 1988) this mode of display has become the most

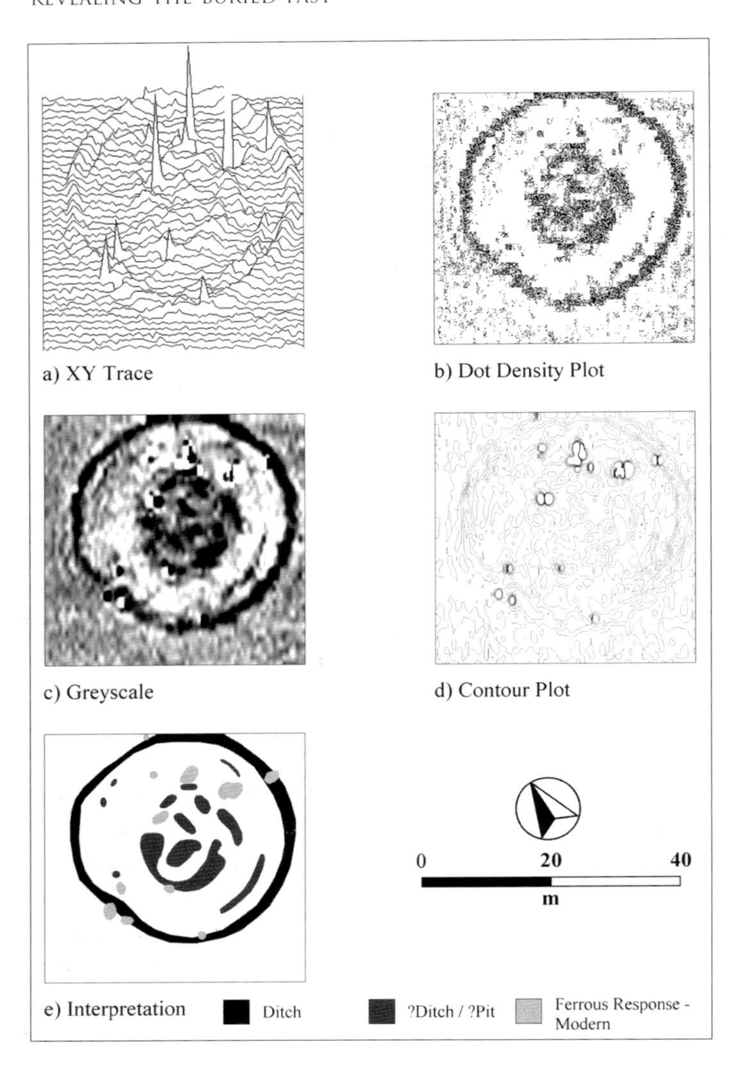

51 *Four types of data display and an interpretation. To understand each display the parameters have to be known. a) 30 nT/cm b) 0.1 to 2 nT c) -2 (white) to +2 nT (black) d) 0.5 nT contour interval. Each display has its merits although it is common to display data in at least two forms for a professional report*

favoured in archaeological geophysics. Increasingly sophisticated monitors and printers allow image reproduction at high resolution and this allows results to be drawn to scale and fine detail revealed. In this format a minimum and maximum cut off is chosen which is used to divide that range of readings into a set number of divisions or classes. These classes have a predefined arrangement of pixels or shades of grey (as opposed to a random number of dots in the dot density plot), and typically the intensity increases with value. The end result is a toned or greyscale image. Similar images can be produced in colour, either using a wide range of colours or by selecting two colours to represent positive and negative values. While colour plots can look impressive and can be used to highlight certain anomalies, greyscales tend to be more informative in general use and are also easier to copy.

The majority of users display datasets with white for the minimum value and black for the maximum. In this way the output mimics the dot-density plots; the paper, or computer screen, is increasingly filled by a darker shade of grey as the value increases. However, the Geophysics Section at English Heritage nearly always produce reverse plots with white equating to the maximum; this is done partly for historical reasons and partly to match the visual effect of ditches as seen on cropmarks. The endpoint tends to be one of personal taste, but obviously it is critical to include a 'vertical' scale bar with any diagram to avoid misinterpretation. While the polarity of black and white is very important, it should also be made clear exactly what the minimum and maximum values that equate to white and black are.

Contours

Although contours are a popular method of displaying topographic variations, the method of display is less common in archaeological geophysics because of the limitations in displaying fine detail in large data sets. Essentially readings of 'equal' value are joined together by a line, drawn in black or a range of colours. Selecting the correct number of contours and the best number of levels can take a lot of practice and quite often several plots may be necessary to investigate particular bands within a broad range of data.

The use of radargrams, time slices, 3D surfaces (**colour plate 13**) and animated gifs may also be appropriate with specific data sets, though in the case of the latter, visualisations on a computer screen cannot be reproduced adequately on paper.

In order to gain a better understanding of the results, it is normally necessary to display the data in more than one format. For example, when assessing magnetic responses it is often possible to differentiate between archaeological and non-archaeological information based on the form and shape of the anomalies. In this case greyscale or dot density plots are used to provide plan information, and profiles or XY traces help in analysing the anomalies, for example to establish whether an anomaly is the product of an iron spike or an archaeological pit.

Whichever display format is chosen, it must be remembered that the display is only a means to an end; the best interpretation, based on theoretical and practical experience, is the ultimate goal. Aesthetic plots do not necessarily make it easier to interpret the data. In fact, misuse of colour in display plots can result in the inexperienced eye giving more emphasis to the results than is warranted.

Interpretation

> Interpretation is an holistic process and its outcome should represent the combined influence of several factors . . .
>
> (David 1995)

The interpretation of archaeological geophysical data is not an exact science as there is interplay between theory and experience. While a broad knowledge of geophysical techniques and the principles of archaeological geophysics are a necessary requirement, other factors are also important. In particular, an appreciation of the nature of archaeological features being investigated is fundamental as is an understanding of the local conditions at the site – including the geology, pedology and topography. It is argued by David (1995) that interpretation should be the work of a qualified specialist with appropriate experience, at least three years in the deployment and interpretation of such surveys. While this is essentially true for professional surveys, that level of expertise is rarely achieved in amateur work. Yet amateur involvement in archaeological geophysics has often been championed as the techniques are non-invasive and, therefore, each survey is repeatable. However, repeating a survey is very tedious and costly; no matter how small scale the survey or how trivial the results may seem, the whole of the data should be fully interpreted and reported on at least to 'archive' level.

Perhaps the most important aspect in interpreting geophysical data is to establish consistent terminology. By this we mean that when an anomaly is characterised it is important that the reader of a report understands what the interpreter means by that classification. Normally the terms are graded by confidence. For example there is an implied decrease in confidence in the following categories:

ditch ⟶ *archaeology* ⟶ *?archaeology*

In such a classification '*ditch*' would only be used when there is clear evidence from another source. Without supporting evidence from an earlier excavation or an aerial photograph the interpretation would be downgraded to '*archaeology*'. If there is no associated geophysical evidence then the anomaly may be classified as '*?archaeology*'.

Depending on the complexity and certainty of the nature of the archaeology at a site it may be possible to group the interpretation of the anomalies in a highly structured manner. In the case of recent work at the Roman city of Wroxeter it was felt possible to reduce the observed magnetic anomalies into 15 distinct classes – see **Table 1**. This level of categorisation is unusual in archaeological geophysics and is the result of the known archaeological context that is supported by a considerable amount of additional archaeological information (Buteux *et al.* 2000).

Opposite:
Table 1 *Anomaly classification used in the interpretation of magnetic data collected at Wroxeter.* After Gaffney *et al.* 2000

Wall (positive): Linear anomaly with increased magnetic signal (*c*.5-7nT), sharply defined edges and forming a rectilinear pattern with other such anomalies. Judged to represent the stone footings of a wall forming part of a Roman building, the stone being more magnetic than the surrounding soil.

Wall (negative): As above but the anomaly exhibits lower magnetic signal than the surrounding background. The stone footing being less magnetic than the surrounding soil.

Road: A broad (*c*.5m) linear area interpreted as representing a Roman road. Either a distinct negative linear anomaly flanked by two positive ditches, or simply a linear absence of anomalies in an area otherwise densely packed with features. Only marked where there is clear geophysical evidence. Isolated stretches on the same alignment are not joined with tentative continuations.

Ditch (positive): Linear anomaly with higher magnetic gradient than the surroundings (*c*.5-7nT). Judged to be a Roman ditch with a fill more magnetic than the surrounding soil.

Ditch (negative): As above, but the anomaly has a lower magnetic gradient than the surrounding soil. Represents a ditch filled with material less magnetic than the surrounding soil.

Defences: Magnetic anomalies associated with the remains of the Roman town defences. Indicates general areas of raised or lowered magnetic gradient caused by the topographic effects of the still extant earthwork remains. Where more specific anomalies can be identified as components of the defence, these are used instead.

Discrete pit (positive): A small area (*c*.1-2+m diameter) of increased magnetic response judged to be caused by a pit-type feature with a fill more magnetic than the surrounding soil.

Pit (negative): As above, but with a lower magnetic response than the surrounding soil, representing a less magnetic fill.

Industrial: Small area, similar in extent to pit-type anomalies but with a very strong positive magnetic gradient (>30nT) judged not to be due to surface iron. Typically caused by the remains of a fired clay structure such as a kiln or furnace. Where associated with stone structures they may indicate the remains of a domestic hearth or hypocaust system.

Disturbed area (structural associations): Archaeological anomalies that appear to represent the remains of a Roman stone building but where no clear building plan can be discerned. Usually identified by a concentration of pit-type anomalies in an approximately rectilinear area. This category is then used to indicate the estimated perimeter of the building.

Disturbed area (archaeological): Denotes an area of disturbance, usually indicated by increased soil noise and judged to be of archaeological significance rather than of modern origin.

Modern disturbed area: Area where the ground has been disturbed in the recent past. Characterized by very large magnetic gradients and a high level of noise often accompanied by concentrations of dipolar, near-surface ferrous responses.

Modern pipe: Straight, linear anomaly with very large magnetic gradients alternating regularly between positive and negative polarity.

Geological: Indicates anomalies of possible geomorphological origin.

Previous excavation?: Area of uniform magnetic signal contained within a well-defined boundary in regions otherwise densely covered with archaeological anomalies.

Background Resistance (R ohms)	Survey/soil conditions
<40	Badly drained, high water table, deep topsoil, springs, boggy areas, adjacent to rivers, clay soils
40-200	Typical rural and urban environments: gardens, grassed areas, fields etc where topsoil is typically 30-40cm deep
200->1000	Thin topsoils, <20cm, dry conditions, very good drainage, sandy and gravely soils

Table 2 *Typical resistance values.* After Walker (1991)

Resistance surveys

As we have seen in a previous chapter, variations in background resistance will tend to reflect the underlying geology and soils. Walker (1991) calculates typical background resistance levels for the scenario of a 0.5m spacing Twin-Probe configuration, with a remote fixed probe separation of 1m (**Table 2**).

Against these generally broad changes will be superimposed smaller fluctuations that are associated with the buried archaeology. Walker suggests that as a general rule of thumb, changes of less than five per cent are attributable to soil noise and changes greater than five per cent may be archaeologically significant, although in practice the observed changes are often much greater.

We have already summarised the types of features likely to be detected with the technique in chapter 2. It has been shown that in general, features that are well-drained compared to their surroundings will result in an increase in resistance, while features that retain moisture produce a decrease in resistance. The problems of reverse response polarity have also been referred to and this is an easy trap to fall into when interpreting results, even for experienced operators. It is often best to establish an estimate of the physical width of a buried feature and this is addressed by analysis of the width of the response, although it is not a 1:1 correlation. A guideline is that if an XY trace is drawn of the resistance over an anomaly, then the width of the anomaly at half of its maximum height is equal to the width of the feature. However, difficulties can arise if two features are close together, as they may result in only one peak rather than two. Other complications occur when considering the strength of response and the depth of the feature. Clearly a feature at greater depth will have less of an effect on an electric current than one closer to the ground surface. But the change in response is not a simple linear relationship. Again Walker suggests that a wall 0.25m below the surface, producing a reading of 100 ohms, might only give a reading of 10 ohms at a depth of 1m.

In addition to archaeological features, changes in resistance will occur where there are variations in topography, vegetation and agricultural practice. A whole range of other man-made features such as pipes, cables and drains will also affect the readings. The change in anomaly strength associated with these non-archaeological features can be on a par with, or much greater than, archaeological responses. It is important that a suitable strategy is devised that will collect sufficient data to establish the likely origin of the anomaly.

Magnetometer surveys

In the case of gradiometers, the interpretation of magnetic data is in many ways easier than for resistance surveys because the problems of a changing background (due to regional gradients or diurnal changes) are in effect stripped out at source. The resultant signals have a bipolar response centred around zero and characteristic negative and positive anomalies arise depending upon the nature of the buried feature.

Clark (1996) neatly shows the effect of both feature size and depth (below the ground surface) on the measured magnetic anomaly using a 0.5m fluxgate instrument, with the bottom sensor being held at 0.3m above the ground. The susceptibility of the pit in both scenarios is assumed to be 100 SI units greater than the surrounding subsoil (**Table 3**).

In fact the strength of the response is only part of the story in that the approximate width of the feature is equivalent to half the maximum signal width. However, it has been suggested this is only true if the depth is less than its width; if the depth is greater than the width then the value should be regarded as an estimate of the depth to the feature (Clark 1996).

A general rule of thumb is that geological features tend to produce broader magnetic anomalies than archaeological features and they produce broader curving anomaly shapes than iron spikes. The variation in response can be quite dramatic and sometimes it is the context that creates a greater confidence in the interpretation.

Distance below ground surface	Pit 1 x 1 x 1m	Pit 1 x 1 x 0.3m
0.3m	4.1nT	2.3nT
0.6m	2.1nT	1.1nT
0.9m	1.2nT	0.6nT
1.2m	0.7nT	0.3nT

Table 3 *Strength of magnetic response.* After Clark 1996

GPR

It is perhaps fair to say that the interpretation of GPR data is even more dependent upon the skill and experience of the operator than other geophysical techniques.

Certain discrete targets will produce characteristic reflections that are readily identifiable. For example, isolated pipes or drums will result in diagnostic hyperbolic shapes, however, in urban environments, the presence of multiple services can produce a mass of reflections that are difficult to untangle. Soil and rock interfaces or voids will result in specific reflections that are recognisable, as such features have all been investigated numerous times and so reference data exists (see Lorenzo *et al.* 2002). The basic difficulty is that unlike pipes and interfaces, archaeological features rarely follow known plans or have specific shapes and forms. Consequently it is difficult to establish reference templates; it is far easier to try to locate known or suspected features and interpret the results rather than to go in blind and try to identify reflections that might be of archaeological interest.

CAD, GIS, and drawing packages

Having decided upon an archaeological interpretation it is important to be able to convey this in a simple graphical way. Although David (1995) claims that in exceptional circumstances, some extremely clear data sets require no further interpretative aid beyond annotation, we do not feel this should be encouraged. No data set really 'speaks for itself', but requires interpretation and explanation. The practice of simply annotating anomalies may lead to false assumptions being made about responses that have not been highlighted. Sometimes apparently clear cut results can be interpreted in different ways depending upon the

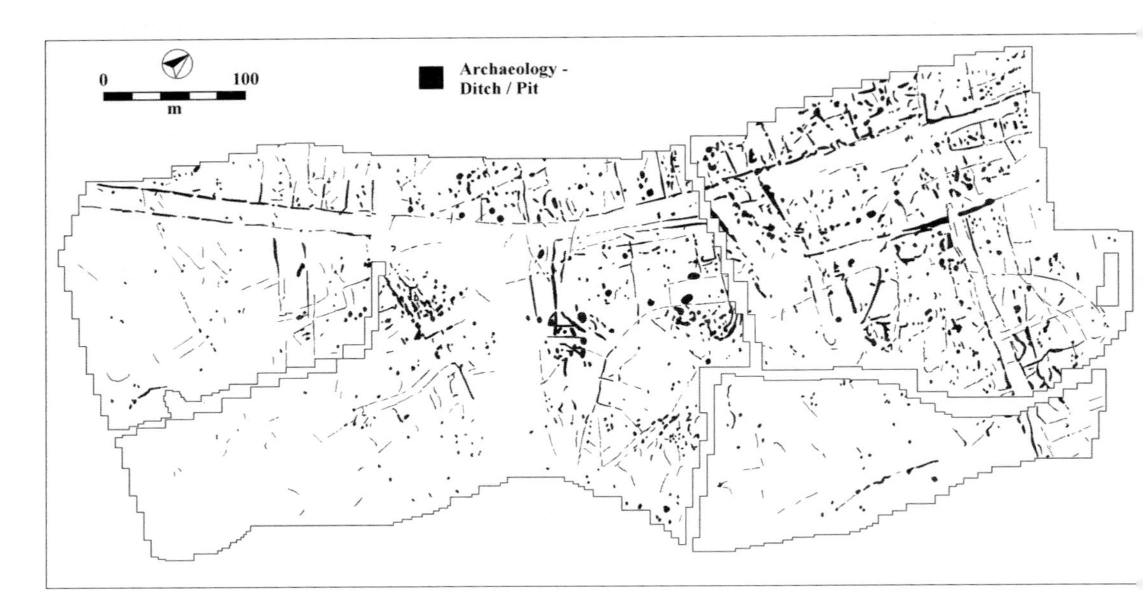

52 *A geophysical survey should always have an interpretation that is separate from the data display. Given the detail in this CAD-based interpretation, from Whitehawk Farm in Kent, it would be completely indecipherable if it was draped over the top of the magnetic data from which it derives.*
GSB for Kent County Council

experience of the investigator, the type of display and the parameters chosen for the display. The interpretation should be reproduced in the report as a separate diagram to the various data displays but, to allow comparisons, they should be at the same scale (**51**).

There is an increasing move towards incorporating geophysical results into vector-based computer programs, such as CAD and GIS. This allows interpretations to be scaled correctly and printed at varying sizes, which is not really feasible if the interpretation has been produced as a bitmap in a drawing program. Digitising an interpretation into a CAD package also allows the information to be used, rather than just viewed, by another group, perhaps one that will use the interpretation to work out an excavation strategy or plan a new building in an area where least archaeological damage can be done. If there is no separate interpretation then this becomes impossible (**52**).

What goes into a report?

Having collected and analysed a data set, the responsibility for properly writing up the study lies firmly with the person or organisation that has undertaken the fieldwork. If the fieldwork initially had to have permission from a national body then there may be a time limit on how long there is to report on the findings.

The report itself will normally be split into a number of headings. David (1995) gives a detailed specification for a geophysical report and this can be summarised as follows:

Title page: Containing information on commissioning body, report reference numbers etc.

Summary of results: A basic, simplified summary of the whole project. The example in **Table 4** is taken from one of our projects and details the information typically included in a database record.

Introduction: Including where the site is (map references), site description, the aims of the work.

Methods: This part should identify which techniques were used and how they were implemented, including information on sampling intervals and data presentation.

Results: This section should describe and interpret the data.

Conclusions: A summary of the results should be related back to the aims identified in the Introduction.

SITE SUMMARY SHEET

GSB SURVEY No	99/55	**NGR**	HY 307 125
SITE NAME	Stones of Stenness	**COUNTY**	Orkney

SITE TYPE Standing stones / henge

DESCRIPTION The monument comprises an outer bank, inner ditch with a break in the north, the remnants of a stone circle and a central hearth-like feature (Ritchie, 1976)

PERIOD Neolithic

GEOLOGY Boulder clays overlying Old Red Sandstone with igneous intrusions

LAND-USE Short grass

SURVEY TYPE	Fluxgate Gradiometer	**METHOD**	Zig-Zag
INSTRUMENT	Geoscan FM36	**SURVEY AREA**	0.8ha
SAMPLE INT	0.25m	**TRAVERSE INT**	1.0m

SUMMARY OF RESULTS

The fluxgate gradiometer survey has confirmed the results of the original 1973 survey by the late Tony Clark and has also added some detail. The nature of the igneous geology is now much clearer. Other anomalies of potential interest, outside of the henge bank and ditch, have been highlighted, though their archaeological interpretation remains speculative. Zones of magnetic enhancement within the ditches may be equated with ritual deposits. Areas of increased noise coincide with material incorporated into the bank during consolidation work carried out by Historic Scotland. Modern blocks marking the location of former stones have also resulted in spurious anomalies.

SURVEY START	8 April 1999	**REPORT DATE**	9 Nov 1999
SURVEY END	8 April 1999		
AUTHOR	J. Gater	**ASSISTANTS**	A. Shields

Table 4 *Example of a site summary sheet from a report*

Plans/plots: Survey location, plots of raw and enhanced data at a variety of scales (always XY traces and greyscales for magnetic data) and separate CAD compatible interpretations.

There should also be information on: bibliographic references; data and interpretation diagrams; appendices; technical information and supporting data such as tie-in information.

Using this template is a good starting point though there is scope for variation and it should not be regarded as the only way to report a survey. Whatever approach is adopted, it is essential that the report must work on two levels: firstly it must stand-alone as technically correct, properly analysed and interpreted and be promptly completed, and secondly, the information must be in a format that will inform and aide other researchers. At no point should the geophysicist lose sight of the wider picture; no matter how important the results may seem, the survey may just be one small cog in a much larger mechanism.

Some examples of survey reports are available on the web at the English Heritage site (www.eng-h.gov.uk/SDB) and in future more should be placed online on the Archaeology Data Service site (ahds.ac.uk).

SECTION 2

CASE STUDIES

Geophysics cannot date archaeological sites or features except perhaps in relative chronological terms. However, archaeologists have been adept at identifying and categorising sites and many 'type' sites can be placed within broad time categories. By making direct comparisons with the results from these investigations, it is possible for the geophysicist to assign similar time divisions to specific datasets.

For the purposes of this book we have divided the case studies into three general categories: prehistoric; early historic; and later historic/modern. It will be seen that some examples are assigned to periods that can be dated more accurately than these categories suggest. This is normally if the survey covers a very particular type of site, such as an Iron Age hillfort or a medieval manor house. Also if the geophysics is tied into pre-existing or subsequent excavation, then specific dating evidence may be available. Many locations that have strategic defensive advantages may be occupied over long periods of time – these are described as multi-period – surveying over such sites can reveal a wealth of seemingly unintelligible anomalies and these can, at best, only be divided into different phases.

6

SURVEYING ON PREHISTORIC SITES

It is clear when delving into the distant past that many sites, or features, defy explanation even when excavated, and given the nature of remotely sensed information this is even more of a problem in the interpretation of geophysical data. However, it is possible to illustrate the use of geophysical techniques on prehistoric sites with reference to the following broad categories: strata definition, palaeochannels, field systems and enclosures, settlement enclosures and settlements, ritual sites and burials. By and large, the archaeology on prehistoric sites is relatively simple, that is compared to deeply stratified urban sites.

Archaeologists often class the features that they find on prehistoric sites as 'negative' archaeology and by this they mean that the features are cut into the subsoil or earlier deposits. As these features often silt up with material of higher magnetic susceptibility from topsoil or settlement refuse, the preferred technique on such sites is magnetic survey. However, the nature of the archaeology associated with the early man severely limits the use of geophysical techniques. Certainly during the Palaeolithic the dispersed and ephemeral nature of the surviving archaeology means that there is little remaining that the techniques can detect. During the Mesolithic, even though longer lived sites with good evidence for fires exist, finding them provides the sort of challenge that most geophysicists would rather pass by. With the exception of locating underground cave systems, see for example Noel and Biwen (1992), Chamberlain *et al.* (2000), Pringle *et al.* (2002), geophysics is unlikely to assist substantially in the search for early hominids or early hunter-gatherers.

However, once such sites have been discovered, often by chance or field-walking, geophysical techniques can be used in a variety of situations to help archaeologists who are investigating Palaeolithic or Mesolithic sites. The most obvious is the use of magnetic techniques to locate areas of burning within a stripped excavation area where there are no visible indicators on the ground surface. In this instance very closely spaced sampling intervals are required and, due to the shallow nature of such deposits, the investigation is usually best undertaken with the Bartington magnetic susceptibility coil.

Strata definition

Other ways in which geophysics can help on early prehistoric sites is in estimating the depth of deposits which overlie bedrock, or in identifying major differing soil horizons. For example, at Cooper's Hole, Cheddar, as part of a *Time Team* investigation into the cave site, it was important to try to establish the thickness of deposits overlying the original cave floor. The information was needed to assess how much material would have to be physically removed before the archaeologically interesting layers were reached. While coring might have provided a simple solution to the problem, and GPR a more sophisticated approach, the low cave roof and clay deposits meant that neither was possible. It was decided instead to employ seismics (**53**). Twenty-four geophones were used at 0.5m separations and spot heights were established using an EDM system. A 12lb hammer and metal plate were used as the energy source. The results indicate that the clay and rubble deposits overlying the bedrock floor decrease in depth further into the cave. At the entrance they are in the order of 3m deep while 12m into the cave they are only 1m deep. A good refracting layer was clearly identified below the deposits; the velocities suggested that this surface equated with the solid stalagmite floor of the cave. Subsequent excavation confirmed these results.

Palaeochannels

Perhaps the main use of geophysics in early prehistoric investigations is in the identifying of palaeochannels or niche-type environments that are likely to have

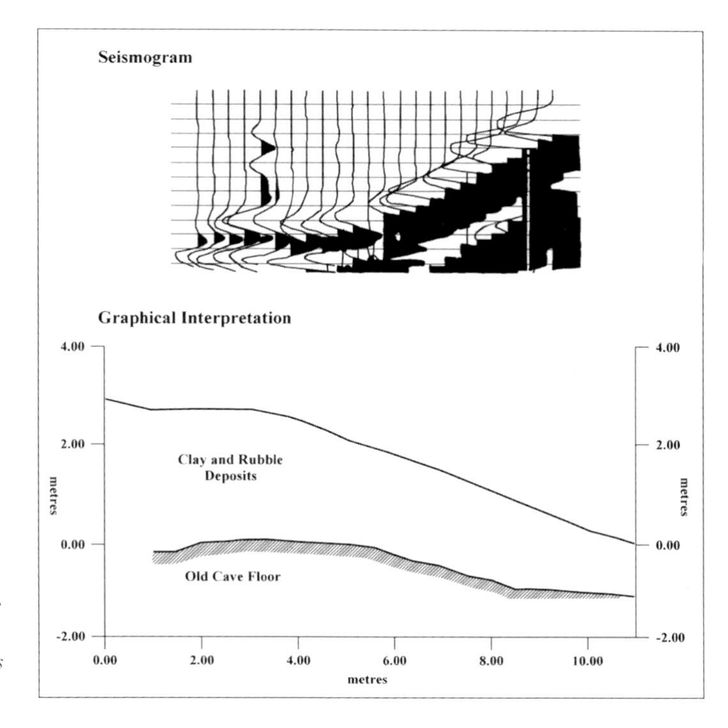

53 *An example of a seismogram which was used to produce an estimate of the depth of deposits overlying the old cave floor at Cooper's Hole, Cheddar*

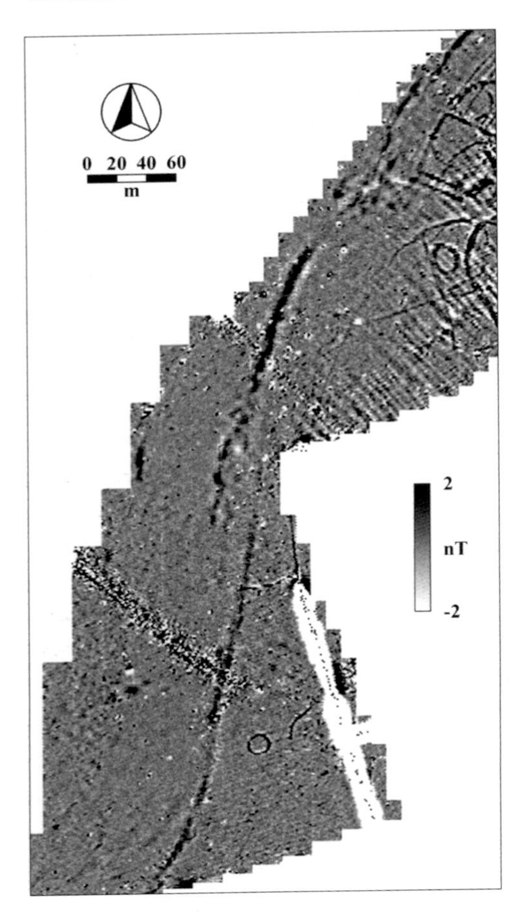

54 *This data set from Riverside Meadows, Bedfordshire, contains a multitude of anomalies from many origins. There is evidently an archaeological 'site' in the north-east corner of the survey, as well as pipes, paths and ploughing trends. What is notable here is the intermittent band of broad and weakly positive response that is the result of a former river channel. Fluxgate gradiometer survey. 1 x 0.5m.* GSB survey for Albion Archaeology

attracted early man as they were obvious sources of both water and wildlife. A paper by Weston (2001) has highlighted the use of prospecting techniques in identifying palaeochannels. While resistance methods, particularly electrical imaging, can play an important role in investigating such channels, magnetic techniques are ideally suited for mapping their courses. At Riverside Meadows we carried out a fluxgate gradiometer survey in advance of proposed development and identified a plethora of magnetic responses that can be equated with ditches and enclosures (**54**). However, in this instance the anomalies of interest are those associated with a palaeochannel that runs parallel to the existing river course. It is very easy to follow the course of this channel even though the magnetic anomalies vary along its length. The bands of magnetic gravels and alluvium that fill the former channel vary according to the speed of flow of the water, the deposition rates and the source material being deposited. As a consequence a small-scale survey may not be able to identify a channel due to a lack of magnetic contrast at a particular point. It may, therefore, be necessary to sample large areas in order to be able to resolve the course of most palaeochannels.

Field systems and enclosures

Field systems are the ubiquitous by-product of an organised, more sedentary life that describes much of later prehistory. These may range from the division of land for stock control or arable production, to larger boundary markers differentiating ownership of land. These can take the form of lynchets, pit alignments, ditches, banks, hedges, fences and walls; each will leave a different geophysical signal when they are removed from the landscape by being dismantled, ploughed out or otherwise destroyed (**55**).

Early fields tend to lack the regularity and complexity of later planned landscapes. Extensive field systems can survive as earthworks or they are often visible from the air as cropmarks. Consequently, since the patterns and alignments are often well documented, templates already exist to compare with the results from a geophysical survey and these help considerably with interpretation. Many of the clearest geophysical images of field systems inevitably come from the later part of prehistory, and the overlap with those from later periods is often apparent.

Perhaps some of the best examples of old field boundaries come from Cornwall, as seen in figure **56**. From a geophysical point of view these features produce very distinctive, yet highly variable responses. In some instances a double positive magnetic anomaly, flanking a negative response, is visible, while

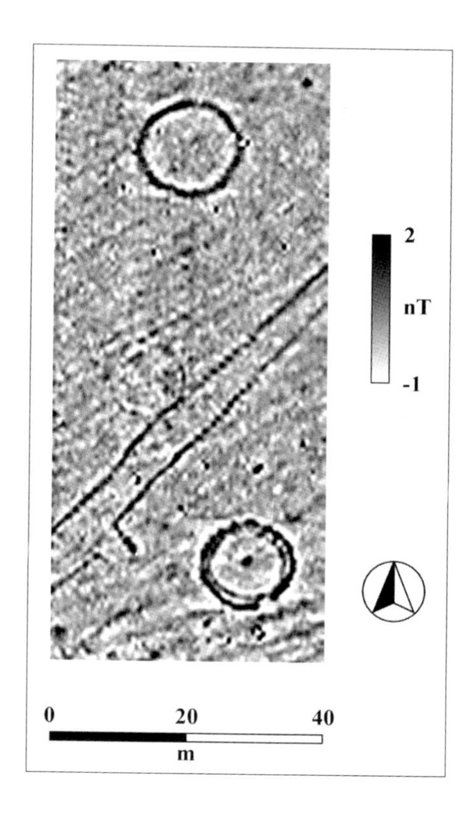

55 *Typical magnetic responses from field systems comprised of ditches cut into the ground. It is impossible to ascertain whether the ring ditches are associated with the former boundaries. Fluxgate gradiometer data. 1m x 0.5m.* GSB survey for CPM

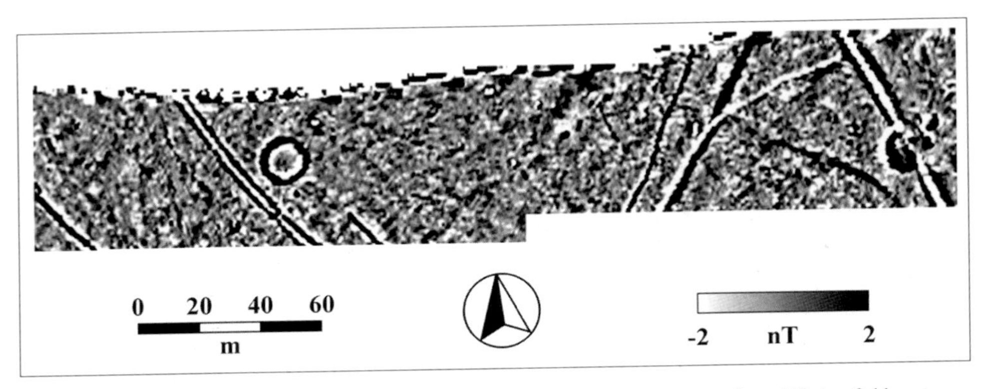

56 *Magnetic survey, from Probus in Cornwall, illustrates the variety of responses from differing field systems. Here we have not only single positive responses over buried ditches, but the typical Cornish double response. Fluxgate gradiometer data. 1 x 0.5m.* GSB survey for Cornwall Archaeological Unit

in others there is either a positive and negative anomaly, or simply a positive response. These responses can be associated with a) ditches flanking a bank; b) a bank with a single ditch; or c) a bank comprising igneous stone faces and an earthen core. It is certain that the dating of any boundary is fraught with problems as in many instances they follow the same line for hundreds, if not thousands of years (**57**).

While geophysics may have limitations in dating archaeological remains, it often comes into its own when investigating cropmark sites and checking the accuracy of any aerial photographic rectification. One prehistoric site studied in the early 1990s near Rugby, Warwickshire, clearly demonstrates the level of errors that can occur, in this instance probably as a result of field boundaries being moved. A maximum discrepancy between the geophysics and the rectified plot of some 20m in east-west and 10m north-south was noted (**58**). Also there was a considerable difference in the shape of some of the enclosures, while others correlated far better. This example does not suggest in any way that aerial photographic rectification 'does not work' or that geophysical data sets are necessarily 'better'; geophysical survey has been used here to refine the aerial photographic evidence and has produced a more accurate and, importantly as the enclosure is no longer skewed, a more meaningful archaeological interpretation.

Settlement enclosures and settlements

Settlement sites, because of their very nature – lots of burnt features and rubbish deposits – invariably produce good magnetic results, even in the prehistoric period. Clearly the range of sites is vast and only a few examples can be included here, but these will hopefully serve to demonstrate the diversity of sites and features that can be mapped geophysically.

In the late Bronze Age one particular type of site associated with habitation that is common in the landscape are burnt mounds. They are scattered throughout

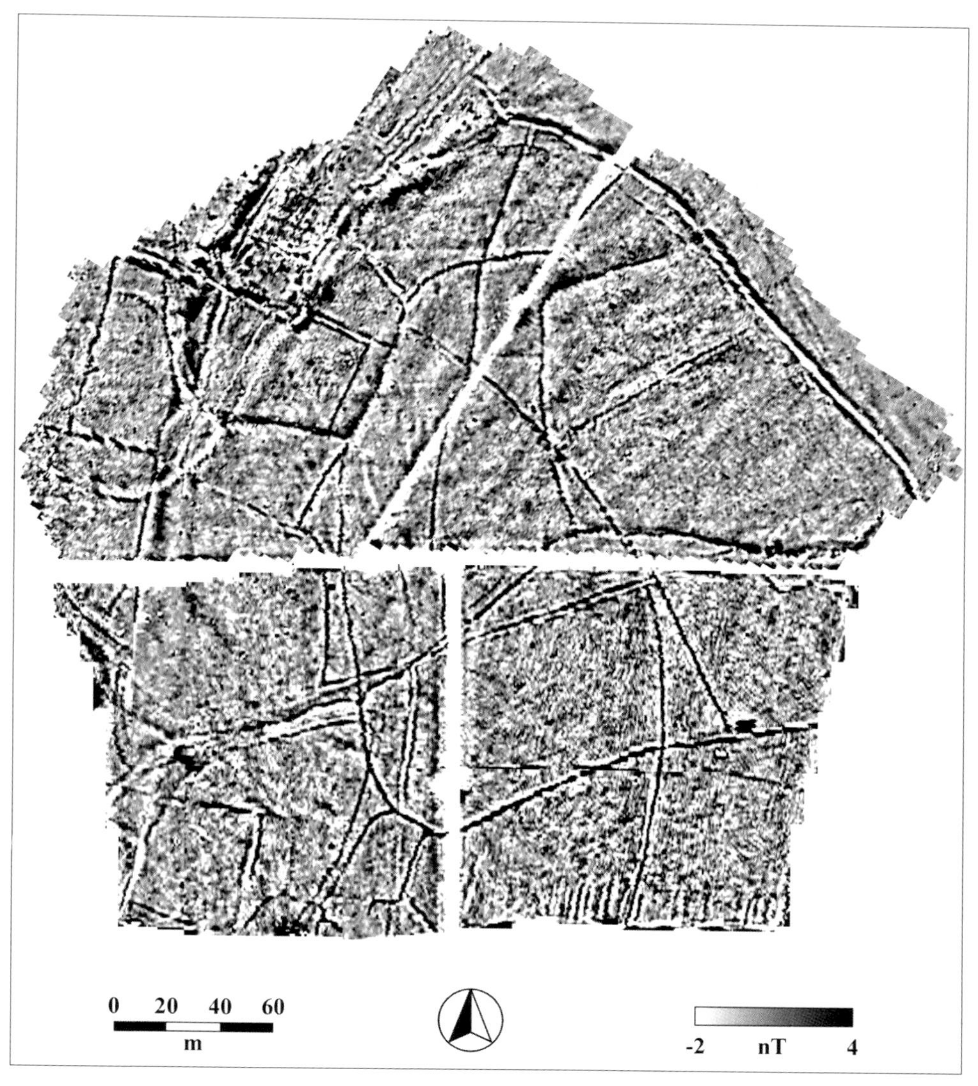

57 *In some areas geophysical survey can reveal field systems that are the product of many thousands of years of landscape management. This survey from Cambourne, Cornwall, illustrates this point; at the north-eastern edge of the survey the present field boundary mirrors an earlier boundary while elsewhere, the modern boundaries, seen here as white 'blanks' cut across the former landscape divisions. Fluxgate gradiometer data. 1.0m x 0.5m.* GSB survey for Cornwall Archaeological Unit

southern England, and extend northwards, with major concentrations in the Isle of Man, Scotland and particularly in the Northern Isles; burnt mounds are also common in many other parts of the world. Although they vary greatly in size and date, they broadly comprise large heaps of fire cracked stones and midden deposits. These often form a horseshoe-shape which extends around a hearth and stone water trough. The stones are heated on the fire, submerged into the trough and

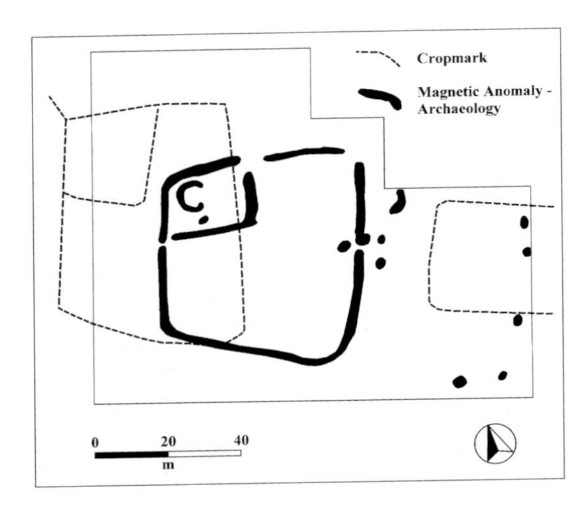

58 *This interpretation shows the spatial differences between a rectified cropmark and area geophysical survey. The interpretation of the latter produces a plan closer to the ground truth. Interpretation produced in a CAD package.* GSB survey for CPM

used to heat the water either for cooking or perhaps as some form of early sauna. The process results in a large numbers of fire-cracked stones that are simply discarded onto the 'burnt' mound. While a few of these sites have been surveyed using other techniques, the more common way to investigate the sites is using magnetic methods. The results from Shelly Knowe in Orkney are particularly dramatic (**59**), in part due to the presence of two adjacent round houses in the data (see Dockrill forthcoming). The responses are particularly strong as features have been dug into the midden material.

During the Iron Age there is a tremendous variety of settlement enclosures (see for example Haselgrove 1999) though one group, so-called 'D-shaped' enclosures, are very distinctive. A good example of such a site is at Norse Road (Dawson and Gaffney 1995) where the geophysics identified a complex of anomalies in and around the 'D' enclosure (**60**). While various alignments are visible suggesting a number of phases of activity, another aspect of these data is of particular interest. The strength of the responses in the centre of the complex is much greater than those further away. This effect has been observed on numerous sites and reflects the decrease in magnetic enhancement of the feature fills on the periphery of the main activity. Known as the 'habitation' effect (see Gaffney *et al.* 2002) this is particularly noticeable when there has been small-scale industrial-type activity or burning in one part of a site and the associated magnetically enhanced material becomes incorporated into nearby open features. Thus the same ditch some 30m away from a source of burning will not have the same level of magnetic response as that close to the activity. While this is essentially a behaviouralist view of rubbish disposal, on other sites the disposal of magnetically enhanced material is potentially more complex (see Stones of Stenness, below).

An unusual group of sites are the so-called 'banjo' enclosures, a descriptive term that derives from their shape. Various archaeological interpretations have been presented, from high-status settlement to stock enclosures, though they have

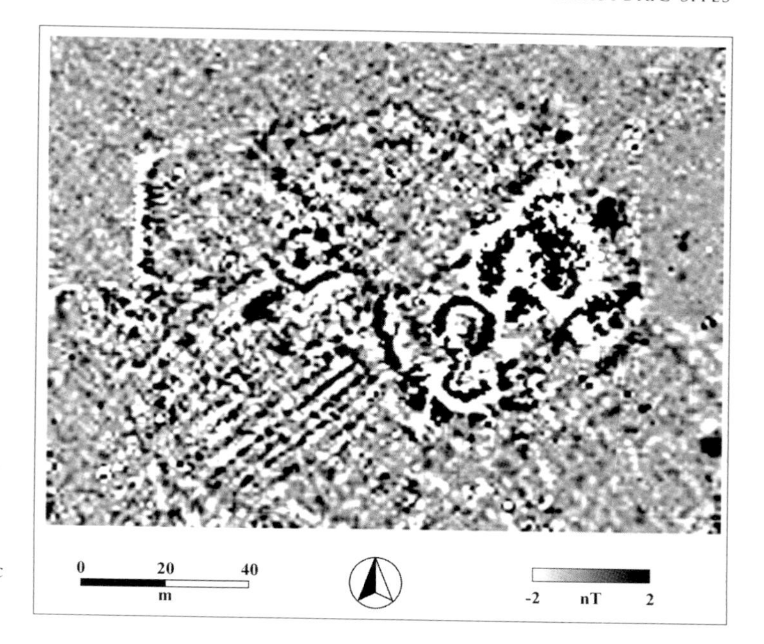

59 *A prehistoric landscape on the Isle of Sanday, Orkney, as revealed by a fluxgate gradiometer survey. The data were hand logged over a 1 x 1m survey grid.* GSB survey for Steve Dockrill/Historic Scotland

also been earmarked as 'ritual' in the past. Three examples are shown; one resistance and two magnetic surveys. The first, the resistance survey at Hamshill, is different to the other two in that it comprises a partially upstanding monument surviving in a densely wooded area. The neck and body of the banjo comprise narrow, stone revetted banks either side of an earthen core, hence the low resistance readings that reflect the latter (**61**). The other two sites are both located in arable fields and have been levelled. The Beach's Barn banjo has a number of large pits inside the body of the feature (**62**) and is surrounded by a complex of other archaeological features, some of a similar period while others are Romano-British

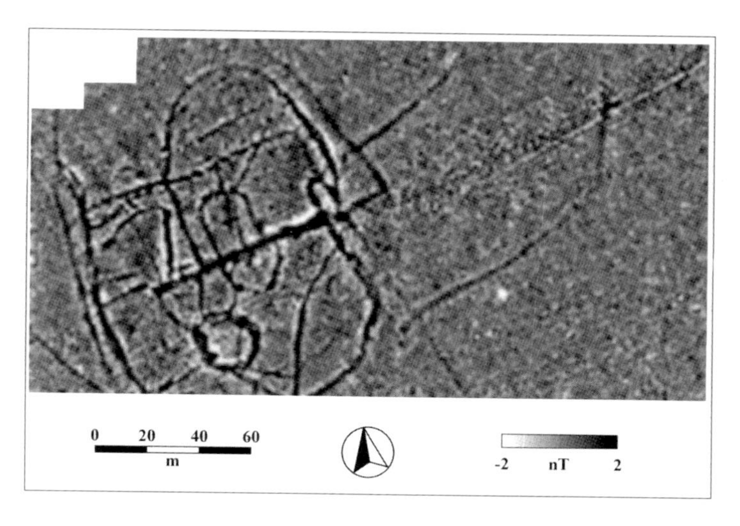

60 *The Iron Age site at Norse Road, Bedfordshire was first noted from the air as a D-shaped enclosure. The magnetic survey is of note due to the weakening of the response away from the core of the site. This decrease is usually referred to as the 'habitation' effect. Fluxgate gradiometer survey. 1 x 0.5m.* GSB survey for Albion Archaeology

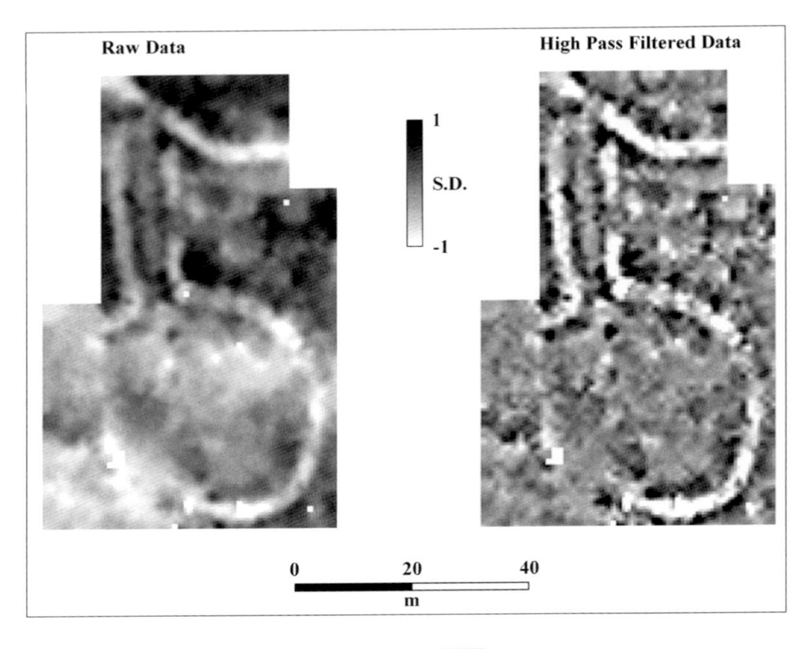

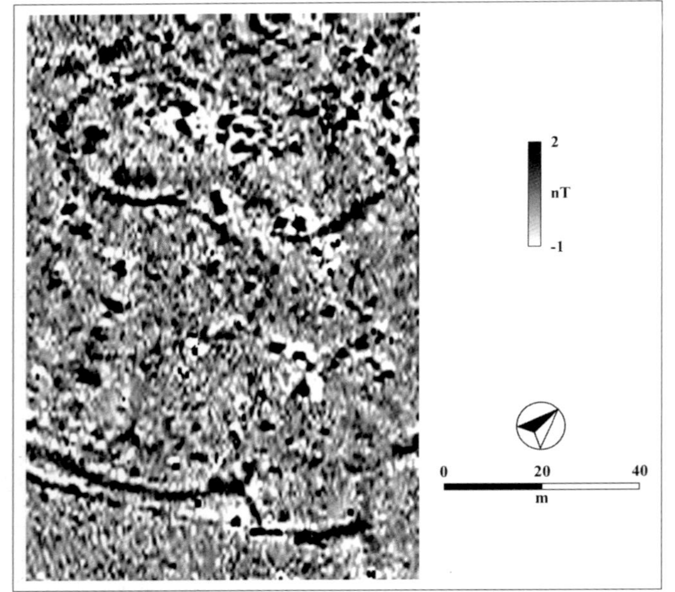

61 (Above) *Twin-Probe resistance survey on a 1 x 1m grid. The raw data have been interpolated to produced a smoother image.* GSB survey for RCHME. © Crown copyright NMR

62 (Left) *At Beach's Barn, Wiltshire, fluxgate gradiometer survey revealed a mass of anomalies suggesting much reuse of the land. Within the data set was evidence for a banjo enclosure. Over the centuries the thin soils over the chalk bedrock give little protection to sites of this type. Data collection 1 x 0.25m; interpolated in image*

in date. By contrast, a banjo near to Leamington Spa appears to be in isolation, a much wider survey failed to identify any other archaeological features, although it contains one large central pit (**63**). The site is also intriguing in that there is a small annex attached to the main body. The variation of internal detail suggests that these sites did have an everyday settlement function rather than some unknown 'ritual' use.

One of the earliest applications of magnetometers was in the study of hillfort interiors (see for example Fowler 1959; Aitken and Tite 1962) where work in the late 1950s, with a proton instrument, was followed by test excavation to verify the findings. Geophysicists are not normally afforded this luxury of being able to immediately excavate anomalies. At sites where a subsequent invasive archaeological investigation does take place, the excavation is usually undertaken some time after the survey and is often carried out by a totally separate organisation. With some of our work, notably the television programme *Time Team*, we have been very fortunate in that small trial excavations are regularly based on the results of geophysics and are often undertaken immediately after the data has been collected and analysed. Thus at Gear hillfort, in Cornwall, where approximately 7ha were surveyed in under three days, we were also able to see the results of excavations within a very short period of time. The complexity and density of the settlement as revealed by the geophysics within the hillfort is quite remarkable (**64**). There are zones of activity, trackways, ditched paddocks, enclosures and numerous round houses visible in the magnetic data. The geophysical evidence was supplemented by carefully targeted trenches which provided critical dating evidence and also an assessment of the state of preservation of the features.

A number of hillfort sites in southern England have been investigated geophysically by English Heritage (Payne 1996). Having well-defined limits, usually in the form of massive defensive ramparts, these sites provide a discrete area which can be studied in their totality. Payne has demonstrated that the results of the geophysical work not only provide valuable information on the nature of occupation at the sites

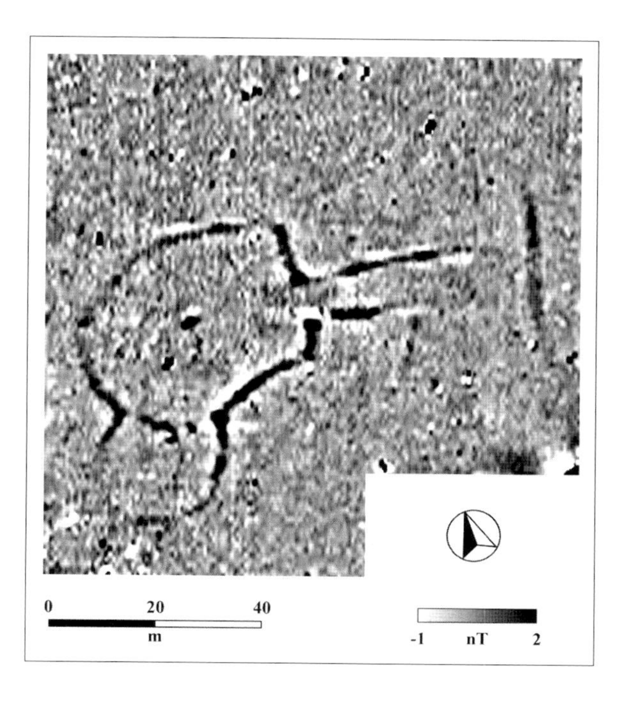

63 *The banjo enclosure shown here has an annex attached to one side. In contrast to the previous two examples this site lies within an area of relatively few archaeological sites. Fluxgate gradiometer survey. 1 x 0.5m.* GSB survey CPM

– the density of dwellings and storage pits, for example – but they can also aid the conservation and management issues at such sites (**65**). While this role of geophysics has become more recognised, the fact that geophysics can help locate and identify new hillforts is perhaps surprising. At Westwood, in Somerset, this is exactly what happened (Gater *et al.* 1993). While investigating the site of a 'lost' Bronze Age barrow on the edge of a prominent hill in Somerset, for the Royal Commission on Historic Monuments (RCHME), two large curving ditch-like anomalies were noted in the magnetic data. Scanning in the adjacent field identified a complex of anomalies and the gradiometer survey was extended. This detailed work revealed a triple ditched (and banked) enclosure encompassing an area of approximately 2ha (**66**). There is an internal subdividing ditch which contains a number of ring ditches

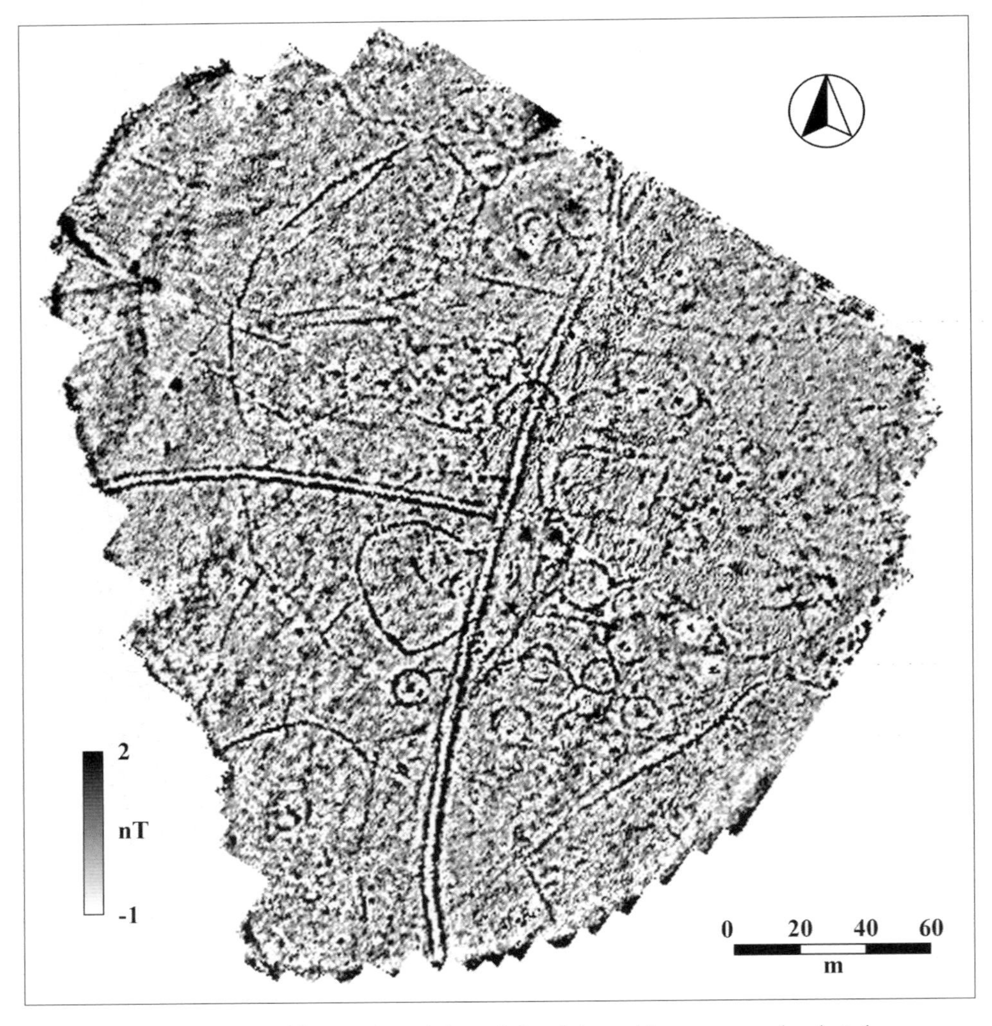

64 *This image is of* c.*7ha of fluxgate data. 1.0m x 0.5m. Prior to this survey no archaeological information, apart from limited fieldwalking evidence, was know from the interior of this hillfort*

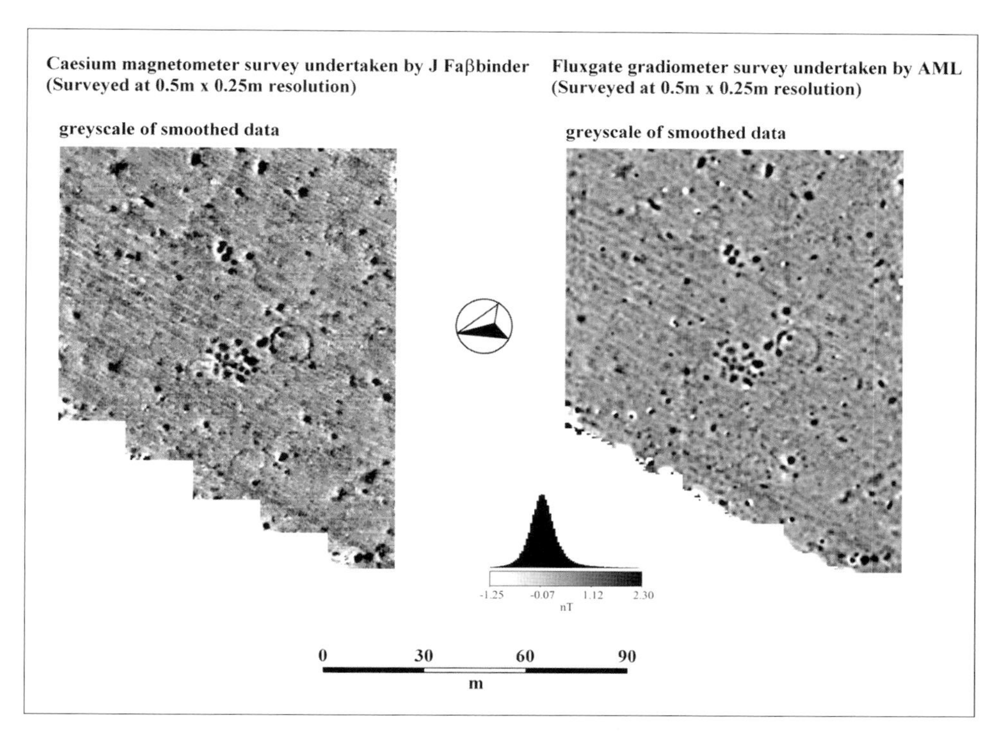

65 *Detailed survey using both caesium vapour and fluxgate gradiometry from the interior of Segsbury hillfort. Given the high resolution of the data collection the images contain much fine detail. As seen elsewhere there is very little difference between the two hand-held devices.* The images were supplied by Andy Payne, English Heritage

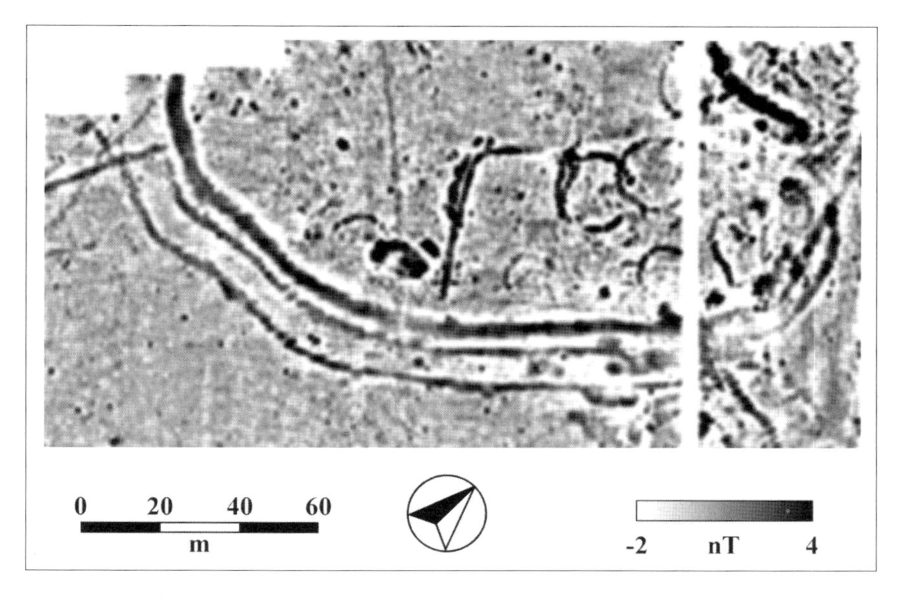

66 *Detailed fluxgate survey (1 x 0.5m) over the 'lost barrow' at Westwood revealed the defences of a hillfort and several ring ditches.* GSB survey for RCHME. © Crown copyright NMR

and an apparently open area presumably used for the penning of animals. Although there is a scatter of pits across the whole of the site, the density is not great, perhaps only 50 pits with diameters greater than 1m in size.

Ritual sites

Depending upon which definition of the word 'ritual' is adopted, the variety of sites that fall into the category is endless. While some types, such as cursus monuments, are archaeologically important, many 'ritual' sites provide little characteristic in terms of geophysical response. Therefore, we will confine ourselves to a few of the more unusual sites where geophysics has revealed important archaeological patterning.

One of the best known, early magnetic surveys that provided a dramatic geophysical image was by Clark (1973) at the stone circle and henge monument known as the Stones of Stenness in Orkney. The magnetic anomalies associated with the ditch produced a striking plot of the site with igneous dykes (strongly magnetic volcanic formations) clipping the edge of the monument. The site was recently re-surveyed and then extended to cover a wider area of the surrounding landscape (GSB 2002). Parts of the resultant plot, reproduced here (**colour plate 14**), again show the ring ditch and the complex of dykes, which could easily be mistaken for field boundaries without proper analysis of the nature and strength of the magnetic anomalies. The variability of the strength of the magnetic response from the circular ditch may be archaeologically important. As there is no 'habitation' effect at work here, the variation in enhanced magnetic material may be significant and relate to ritual disposal of that material.

One of our most surprising results was the discovery of a hitherto unexpected ring ditch on a sports playing field in Worcestershire as part of a standard archaeological evaluation exercise. Playing fields are often difficult for geophysics because of a number of complicating factors: past landscaping to level the fields; underlying field drains and service trenches; sand pits and artificial surfaces for practice areas; and magnetic interference associated with metal goal posts. Examples of all these effects are visible in figure **67** but in amongst all the noise is a distinctive ring anomaly measuring approximately 30m in diameter. Unfortunately the north-eastern arc of the ring is magnetically disturbed by modern ferrous material, thus it was not possible to be certain whether or not there was a genuine break at this point. Excavation confirmed the existence of an 'entrance' and provided dating evidence to suggest that the site is a henge-type monument.

Another example of a henge, this time a Class II henge, that is a henge with two entrances, is from Catwick in Yorkshire (**68**). The entrances show clearly in the magnetic data and there are a number of internal responses of archaeological potential. However, when the area outside of the henge is also analysed there are similar magnetic anomalies, and this wider perspective makes the interpretation of the internal detail less certain. It is apparent that there are a number of amorphous anomalies across the survey area and they could all be either archaeological in origin or the result of a natural soil effect. By way of contrast, the results from Stanton Drew

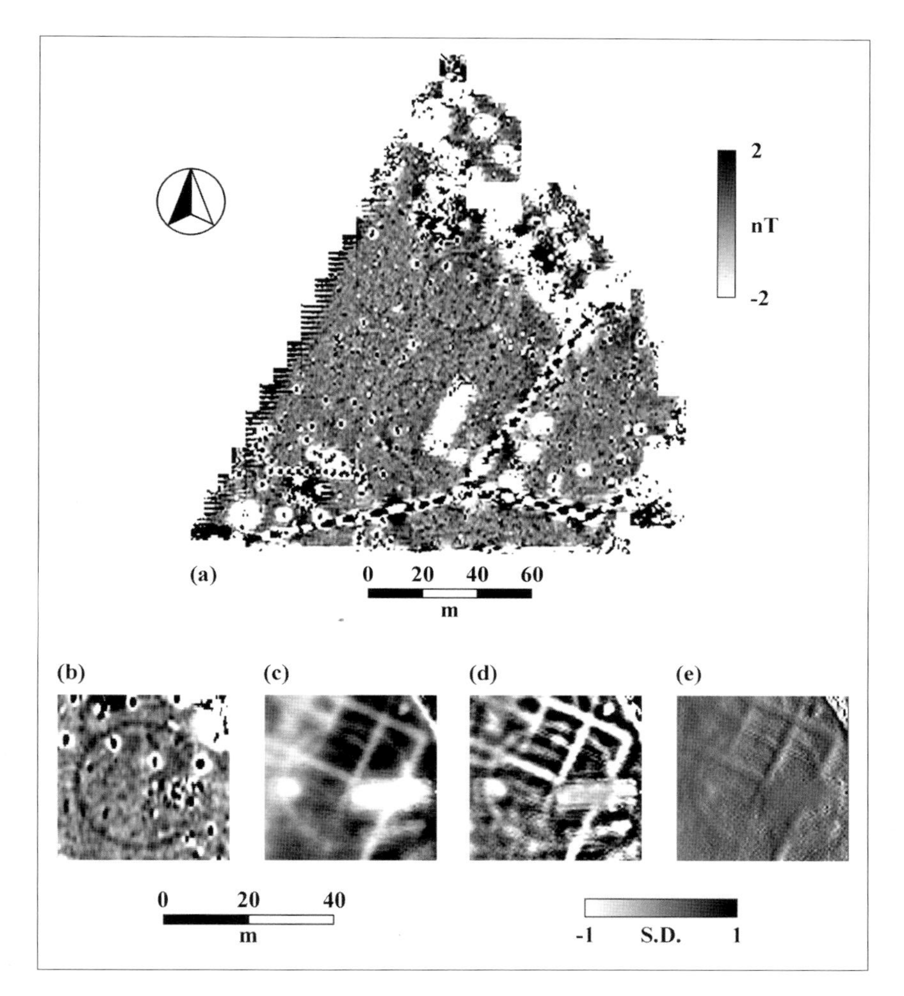

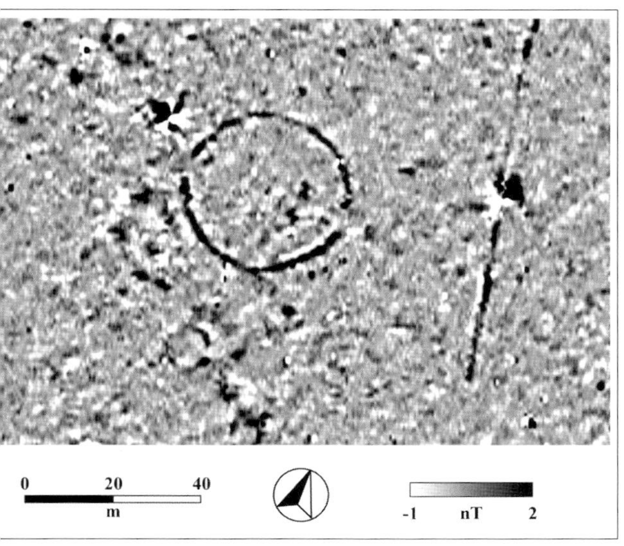

67 (Above) *(a) Fluxgate gradiometer survey over a school playing field at Perdiswell produced many anomalies of modern origin. In amongst the ferrous debris can be seen a ring ditch. After a small trial excavation the ring was resurveyed (b) and also subjected to Twin-Probe resistance survey. The resistance data is shown in (c) raw (d) filtered and (e) relief plot. The large area of disturbance is due to the trial trench, while the grid is a result of white lining on the playing field.* GSB survey for WHEAS

68 (Left) *Fluxgate data set 1 x 0.5m. A Class II henge from Catwick in East Yorkshire. Whilst there are many anomalies of possible interest within the henge, similar anomalies can be found outside of it; it is possible that all of the pit-type anomalies are natural rather than the product of anthropogenic action.* GSB survey for ARCUS

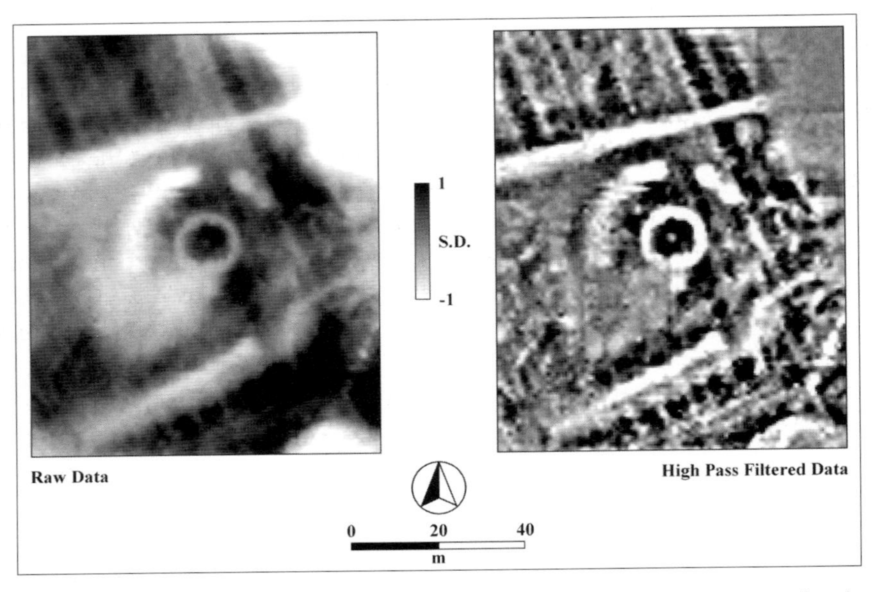

69 *The Twin-Probe resistance data set. 1 x 1m. The data on this henge-type site were collected by Continuing Education students undertaking a short course in archaeological geophysics run by the University of Leeds in conjunction with GSB*

in chapter 3 show the amazing internal detail that can, under good conditions and high sample density, be extracted from henge monuments (see **31**).

Also probably falling into the category of henge is a site first noted from the air in North Yorkshire. Magnetic surveying at the site near Gargrave hinted at the complexity of the site; however, the responses were weak due to a lack of magnetic enhancement that is typical on an infrequently used site. Fortunately, the resistance results have provided an unambiguous plot indicating a central circular low resistance ditch anomaly measuring 15m in diameter (**69**). There may be a central pit and a possible area of high resistance in the inner north-east quadrant. Beyond the north-west and north-east extremities of the ring are two arcs of low resistance readings suggestive of ditches. There is no evidence for matching arcs in the south-west or south-east. Two straight linear ditches run to the north and south of the site, though whether an integral part of the monument or later field boundaries is uncertain. The diagonal trends crossing the survey area are associated with ploughing, some of which seem to stop at these 'boundaries', although others continue through. Even though the results are clear, without excavation the interpretation of this collection of responses will remain puzzling.

A totally enigmatic site which provided perhaps even more dramatic and unexpected results is Mine Howe in Orkney. Here, a series of steps was rediscovered by a local farmer extending some 7m down into the centre of a prominent earthwork mound. During a detailed geophysical investigation, which was carried out primarily to try to establish whether there were any chambers surviving in the mound, a substantial ditch up to 8m wide and 2m deep was discovered. Viewed best in the magnetic data (**70**) the ditch encircles the mound with an opening/entrance in the north-west. The complex of magnetic responses at this point is associated with a concentration of archaeological activity, including small-scale metalworking processes.

GPR survey was carried out at Mine Howe but failed to penetrate through the Boulder Clay and so the question of whether there are further chambers in the mound remains unresolved. However, the technique initially provided useful profiles and depth information for the ditch (**colour plate 15**) and aided the excavations. Subsequent time-slicing of the data resulted in a clearer picture of the ring ditch in plan view and at differing depths (**colour plate 16**).

Finally, mention must be made of Stonehenge, which is part of one of the most important archaeological landscapes in Western Europe. It is perhaps surprising that the first geophysical survey of the monument itself was only carried out in 1994. However, during the past decade the land surrounding the henge has been subjected to an unprecedented amount of archaeological geophysics. While some of the work has been research orientated, even more surveying has been undertaken due to the evaluation of land as a result of proposed improvements in the local road network and a suggested new visitor centre for the monument. A result of this is that data have been collected over many hundreds of hectares and it is clear that at this scale the geophysical information is contributing to the understanding of Stonehenge at the landscape level, rather than at the monument, or site, level (David and Payne 1997).

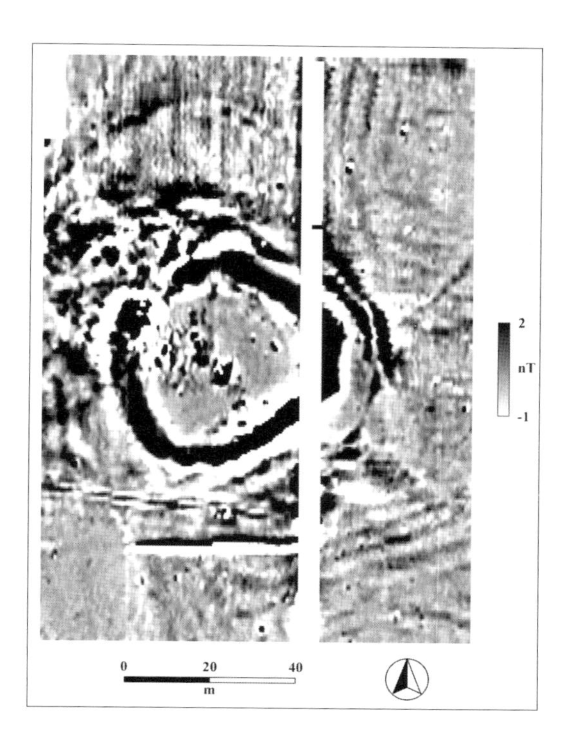

70 *Fluxgate gradiometer data from Mine Howe in Orkney. 1 x 0.25m. The results indicate a substantial ditch surrounding the mound and a plethora of other archaeological responses, particularly concentrated around the entrance in the north-west.* GSB survey for OAT/Historic Scotland

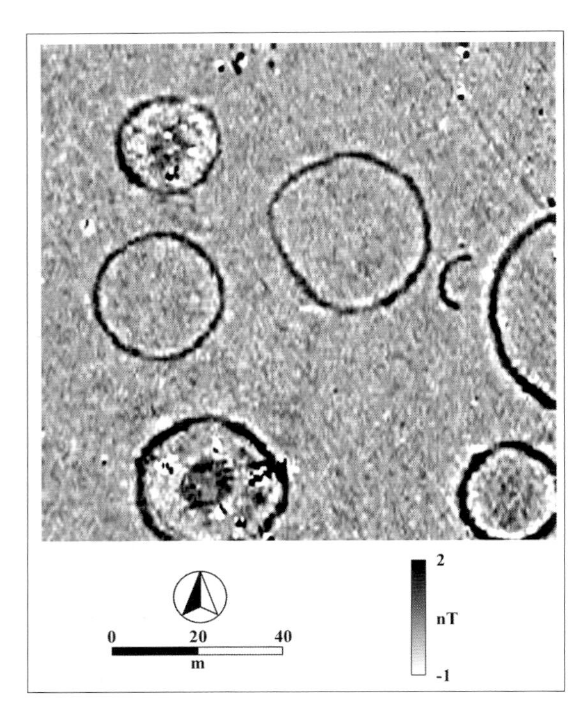

71 *A remarkable complex of barrow ditches, some exhibiting internal features, presumably graves. Fluxgate gradiometer data. 1 x 0.25m.* GSB survey, data courtesy of Dr Charly French, Cambridge University

Burials

The range of funerary practices and burial types in the prehistoric period is immense. While geophysics can play an important role in identifying and mapping many of these sites, it is important to point out that the actual burial, i.e. the body or the grave cut, is a very difficult target to locate. There is no instrument that can detect bodies *per se* in the same way that a metal detector can find metal. Apart from the corpse and any grave goods, normally the same soil goes back into a grave as was dug out in the first place; consequently there tends to be little or no physical contrast that can be detected. However, the burials are usually located within a larger monument and it is these features, such as a surrounding ring ditch, that are more susceptible to detection.

Perhaps the most common burial feature visible in the British landscape, either as upstanding monuments or as ploughed-out cropmarks, are barrows. These come in a variety of shapes and forms but geophysically they tend to comprise of the ditches that surrounded the long or round barrows. Occasionally if fire has been part of the burial practice, an internal burial can be located or if the grave is cut into chalk there may be a measurable magnetic contrast. At Wyke Down in Dorset, Dr C. French of Cambridge University has been carrying out a research project on the barrow group and the surrounding rich archaeological landscape (French *et al.* forthcoming). Part of the research design for the project has involved employing geophysics on a variety of ploughed-out sites and upstanding funerary monuments. While both have produced very clear results, there are interesting differences between the magnetic responses (**71**). Within the surviving mounds there is evidence for internal ditches and pits, presumably graves. Where the sites were ploughed flat, either the ring ditches were the only surviving feature or they contained areas of increased magnetic responses due to enhanced deposits being disturbed and dispersed by ploughing.

A rarer type of funerary site, a Neolithic mortuary enclosure, was identified during the survey of a Romano-British villa complex in Berkshire. It transpired that the villa was destroyed during the construction of a railway line, but during the search for the building the extended magnetic survey recorded a small rectilinear ditched enclosure (**72**). The shape and orientation of the features when compared to the other linears suggested that it dated to a different period from the Romano-British remains. On typological grounds alone it was thought to be perhaps Neolithic and excavation confirmed this hypothesis, though no actual burials were discovered. However, at another location the significance of this type of site could be easily mistaken, especially if the data set included many inter-cutting features.

The construction of the Bedford bypass in the early 1990s provided a rare opportunity to carry out an extensive geophysical survey over a landscape rich in a variety of archaeological features, largely Neolithic/Bronze Age in date. The work was carried out in conjunction with the Bedfordshire Archaeology Service (now Albion Archaeology). In particular part of the area that needed to be

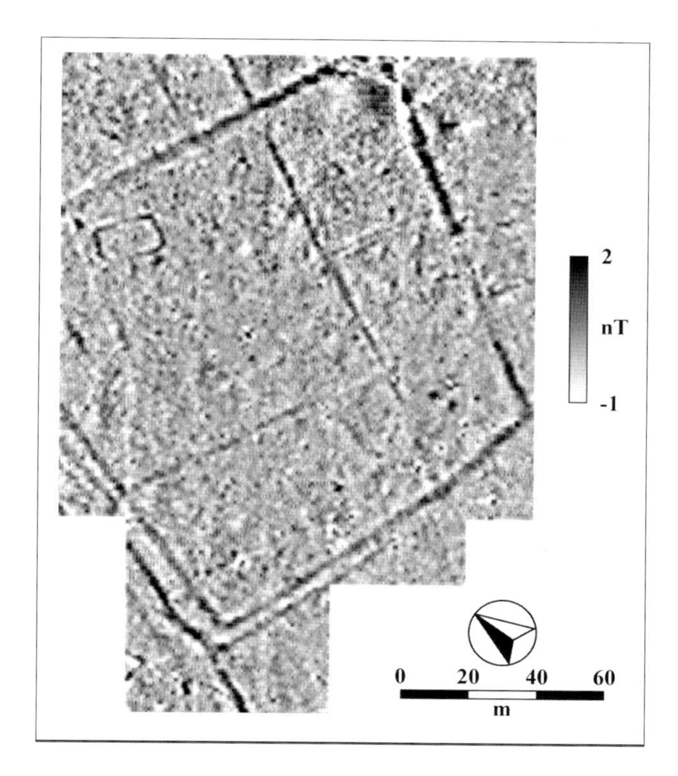

72 *A small rectilinear enclosure within a complex of Romano-British field systems. Excavation suggested that the feature may be Neolithic in date and possibly a mortuary enclosure. Fluxgate gradiometer survey. 1 x 0.5m*

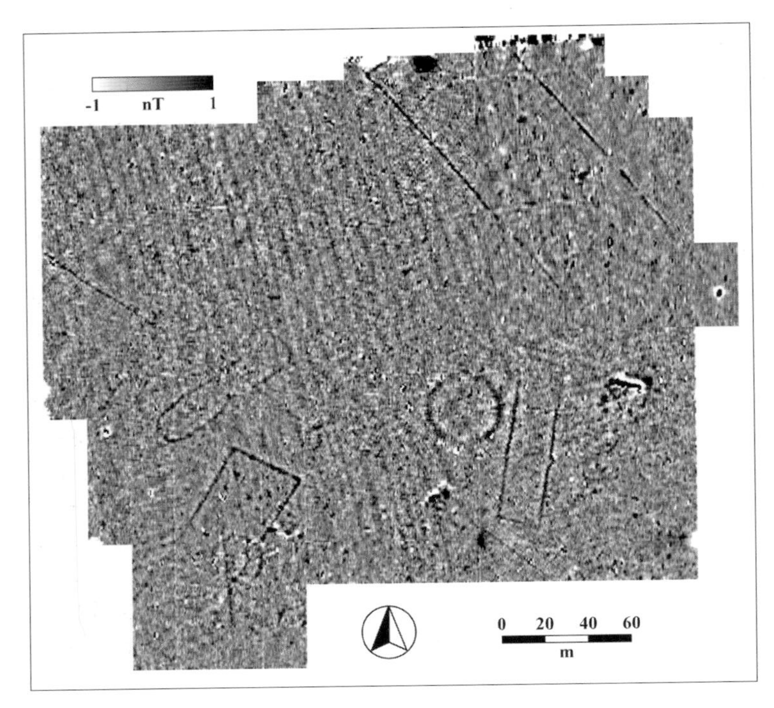

73 *A prehistoric landscape known as the Cardington Cursus Complex in Bedfordshire. Fluxgate gradiometer survey. 1 x 0.5m.* GSB survey for Albion Archaeology

surveyed is known as the Cardington Cursus Complex (**73**). The complex comprises a group of monuments associated with a small cursus; a variety of recti-linear enclosures exist, together with a 'paper clip' enclosure, and they follow a similar orientation to the cursus. There are also a number of circular features in the landscape. The complex is interpreted as a funerary landscape of some considerable longevity. Other complexes have been identified in the Midlands and East Anglia and although limited excavation has been carried out, the monuments are seen as coherent groups and not an accidental grouping. The alignment of the features is seen to be associated with sunrises and sunsets; as such it is believed that the complexes were involved with rites that accompanied the funerary functions of the monuments.

The complex was first identified from the air and fieldwalking recovered a number of artefacts, mainly flint tools and flakes. The aim of the geophysical survey was to accurately locate on the ground the cropmark features and to see if other features invisible to aerial photography could be detected. The site lies in a flat river valley on river sands and gravels, not the best of soils for magnetic enhancement. Furthermore, as we have seen, the nature of funerary and ritual complexes rarely leads to a significant enhancement of the magnetic properties of the soils. In this instance little magnetic contrast was observed between the archaeological features and the subsoil. The ploughed nature of the ground, including areas of potato furrows, added to the levels of noise to the extent that in places some of the anomalies associated with the archaeological features seem to disappear. Thus at one end an enclosure may be clearly visible as a low strength magnetic anomaly, while at the other it is poorly defined. Many of the linear anomalies also have the appearance of being fragmented. This is in part a function of the angle of strike of the collection traverse to the feature, an aliasing effect, but also due to the variable depth of topsoil.

When the results are compared with the aerial photographic transcription there are only minor differences in the plotted locations, though the 'paper clip' enclosure proved to have a discrepancy of a few metres. No major additional archaeological features were located in addition to those identified on the aerial photographs, though a few previously unrecorded linear features were interpreted. However, when the magnetic contrasts are so weak it is difficult to be confident about locating pit-type features, and although a few potential targets were highlighted it is quite possible that others were missed. This is a limitation of evaluation surveys. They tend not to allow for more intensively sampled surveys to be carried out, or for the work to be performed at a time when the ground conditions are most suitable. Thus, even though this is an excellent survey undertaken over a decade ago, the confidence levels remain low when one considers the problems associated with the mapping of subtle, but important, elements of this archaeological landscape, e.g. pits, slots and gulleys.

Wider perspective

Despite the fact that the first ever use of resistance survey in Britain was to try to locate pits of prehistoric date at Dorchester, it is now known that the technique is not very efficient at locating such features. But there are clearly specific cases on early sites where the technique has advantages over magnetometry. An example of one use of resistivity within a prehistoric study is connected with the estimation of the extent of flint mining in Poland. Herbich (1993) provides a thought-provoking set of electrical results using a variation on the 'Schlumberger' array at the sites of Polany and Krzemionki. Initially he performed a series of symmetrical Schlumberger soundings along a line and chose two separations for area survey that would illustrate both the near surface and the base of a known rubble horizon. It was clear that there was a substantial zone in both separations where the resistivities were common. Herbich argued that this was a result of the mixing of the geological layers by excavation and that the near surface variation was the result of undisturbed layers. Furthermore he predicted that the shafts themselves would be in the crossover area between the two zones and this was confirmed on excavation.

Most of the case studies presented in this chapter have shown that magnetometry, in the form of fluxgate or caesium vapour gradiometry, is the favoured approach on the majority of prehistoric sites in Britain. The magnetic technique has also proved extremely successful in the Americas, as can be seen in the spectacular magnetic map from Big Hidatsa Village (Weymouth 1986). However, it

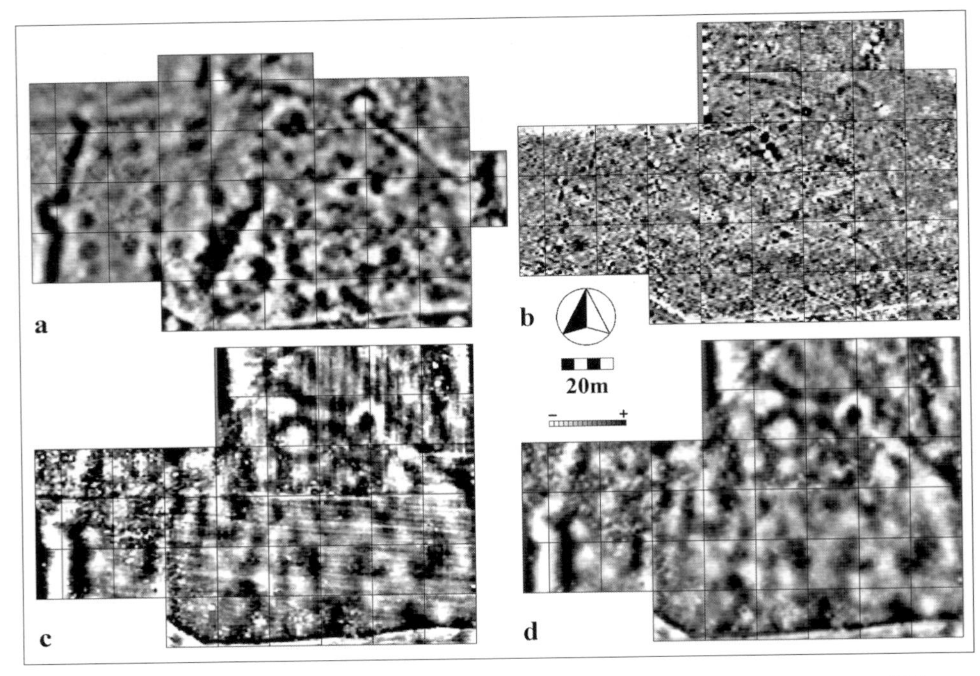

74 *Data from Whistling Elk, South Dakota, USA. The resistivity data (a), collected using a 1m Twin-Probe separation, has illustrated the major elements of this native American village. The gradiometer data (b) is less informative but suggests that at least some of the houses were burned to the ground. The images in (c) and (d) are conductivity data collected using a Geonics EM38. The strong plough marks in the raw conductivity data (c) have been removed in (d) using Fourier methods.* The image is courtesy of Dr K. Kvamme. For an extensive description of the data from this site see Kvamme 2003

has been found that much closer sampling intensities are often required on these native American sites and the same is also true on many European sites in countries like Austria and Germany. In many cases eight samples per metre are often collected along traverses that are separated by 0.5m. The major reason for such high sample density is that the magnetic signal-to-noise is often very low and the target features themselves are small and/or insubstantial. As a result it is important to get a very good estimate of the background in order to work out what is signal rather than noise.

An excellent, recently published example of a multi-technique approach comes from the work of Kvamme (2003) on sites on the North American Great Plains of the Dakotas. Figure **74** shows some of his data collected at Whistling Elk Village which dates from *c.*AD 1300, but in this context a prehistoric site. Although on the majority of sites that Kvamme has surveyed, magnetic survey has proved most fruitful, the increased depth, *c.*1m of overburden, at Whistling Elk Village proved better for Twin-Probe resistance and EM38 conductivity surveying. His work shows that it would be wrong to discount other techniques that are available to archaeologists working on prehistoric sites.

Earlier in this chapter GPR was shown to be used on textually homogeneous soils to detect features that were otherwise invisible to geophysical techniques. Another EM application is over features of greater size. Frohlich and Lancaster (1986) describe the use of a Geonics EM31 instrument to identify shaft tombs in countries situated in the Middle East. This particular EM induction meter works at a frequency of 9.8kHz and has the coils separated by a fixed distance of 3.66m. Used in its vertical format this instrument can induce electrical currents in the ground down to a depth of 6m. At Bab edh-Dhra in Jordan the authors reported discriminating in-filled shafts and tombs and non-filled tombs. A significant interpretation was that while most shafts showed an increase of about five per cent in conductivity, those that had been recently robbed produced much higher increases, perhaps as much as 100 per cent. This of course allows efficient excavation but also significant mitigation strategies to be put in place as the burials can only be efficiently protected if their location is known.

Summary points

• Magnetometry is the preferred technique for identifying near surface cut, or negative, features that are commonly found on prehistoric sites.

• On sites that produce low-level responses it is imperative that a systematic approach is used to produce high quality data.

• 'Traditional' sampling intervals of 1 x 1m are not sufficient for collecting magnetic data on these sites. At least two samples per metre must be collected in one direction, and in some instances, mapping early sites in USA for example, the sample density may have to be increased several times.

• Many settlement sites can be located by rapid assessment using a fluxgate gradiometer and, if conditions permit, magnetic susceptibility can also be used. Ephemeral, or ritual, sites are often difficult to find with any rapid assessment or, indeed, with detailed survey.

• Other techniques should not be discounted. This is especially true of sites that are at a depth beyond the detection limits of a gradiometer.

7

SURVEYING ON EARLY HISTORIC SITES

It is probably true that more geophysical surveys in the UK have been carried out on sites from this time period compared to all the others. Of course there are a great number of Roman sites but the overriding fact is that geophysical techniques tend to work well on archaeological remains from this period. The nature of the archaeology is such that the sites are liable either to have good magnetic enhancement that is ideal for gradiometer survey or, due to the increased use of stone as a building material, they are favourable for resistance and GPR surveys. Although data collected over Roman sites often show considerable complexity, they frequently conform to highly regular patterning and this can lead to particularly good, if not 'true', archaeological images.

While there are many different types of sites within the early historic period, it is convenient to illustrate the potential of geophysical techniques in the following broad categories: roads, field systems, villas, settlements and towns, fortified sites and industrial sites.

Roads

An integral part of any landscape is the network of roads and in the Roman period in Britain roads were constructed to serve both military and supply purposes. They are theoretically simple targets; the roads are often described as substantial stone features, constructed with several layers of hardcore and flanked by a ditch either side. In practice this 'standard' is rarely seen, and there are few geophysical examples; the 'standard' road may be a reality in, or close to, urban or fortified areas – they are clearly less substantial features in more rural locations and areas away from sources of suitable hardcore. Even when the road is classically made the geophysical evidence can be surprisingly slight. In the example shown overleaf, from Cheshunt in Hertfordshire, we were initially confused by the resistance data collected over what was believed to be a section of Ermine Street (**75**). The survey was started at the southern end of a field where the road was thought to emerge.

Within the data a clear high resistance anomaly was identified and we automatically assumed this to be the road surface. The survey was extended northwards and a second linear high resistance anomaly was discovered, aligned parallel to the first. As we expanded the survey, the two anomalies continued to follow each other and then we realised that these were the ditches flanking the road, being set some 17m apart. Our interpretation was that the ditch fills were better drained than the surrounding subsoil, hence the high resistance response, and that the original road surface must have been ploughed away. Excavation confirmed this interpretation; the side ditches had become filled with the gravel that had originally formed the road surface.

Given that Roman roads are justly famous for following straight lines, it is easy to devise strategies for locating them, as long as it is known in roughly which direction they are aligned. A standard procedure is to use search transects that are surveyed at right angles to the projected course of the road, either with magnetometry to locate any side ditches, or with resistance survey to locate the core of the road (see figure **47**). An alternative but similar strategy is to utilise an Electrical Imaging system to investigate a vertical section across the supposed line of the road. A survey of this type was undertaken at Lambeth Palace in the heart of London in an effort to find a Roman road that was apparently exposed in the 1930s. As the road was believed to be buried at about 2m depth, and adjacent to a buried garden wall and a medieval moat, it was decided that standard area survey may not be applicable. As a consequence, imaging data was collected perpendicular to the orientation of the road; the spacing between the probes was 1m and it was anticipated that the current would penetrate about 3m into the ground. The results (**colour plate 17**) show a zone of low resistivity data associated with the moat, a discontinuity linked with the garden wall and a lozenge of high resistivity material. The latter was in the correct location and at the approximate depth of the presumed road. Looking at this highly resistive body it is tempting to interpret the anomaly as the road, but it does not appear to have side ditches or a ground surface associated with it. It was with some reservation that excavation was undertaken at this spot and that caution was justified; a lens of natural gravel, probably fluvial, was unearthed at a depth of *c.*1.3m. This is a good example illustrating that not all anomalies have to be archaeological in origin even though on face value they appear to be.

Field systems

As noted in the previous chapter, field systems are notoriously difficult to date. Even the excavation of these features can often produce little or poor dating evidence and this is especially true when considering managed landscapes at some distance from settlement foci. It is impossible to date field systems in anything other than a broad brush manner using geophysical techniques, and our illustration here is linked to a site that has significant elements that are known to be

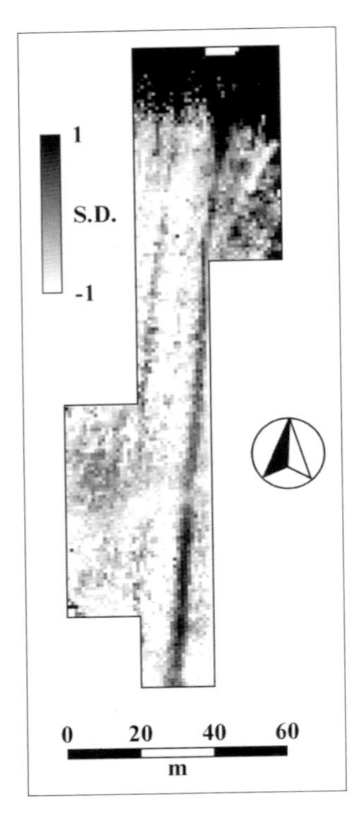

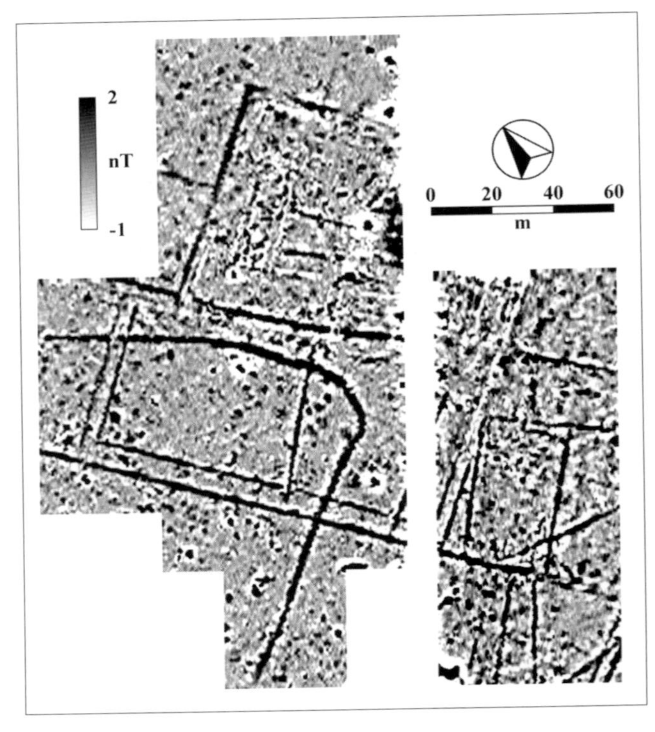

75 *Ermine Street, Hertfordshire. Twin-Probe resistance data. 1 x 1m. The two parallel linear high resistance responses indicate the gravel-filled ditches flanking the road, which itself has been ploughed away*

76 *Waltham Villa, Gloucestershire. The fluxgate gradiometer data, collected at 1 x 0.5m intervals, indicate a complex of Romano-British features associated with the villa building. The main building is situated between the nT scale and the north arrow. In addition, a large curving anomaly was recorded that excavation dated to the Iron Age*

Roman in date. At Waltham Villa in Gloucestershire the planned rectilinear fields, paddocks and trackways associated with the villa complex are clearly visible in the magnetic data (**76**). A separate, large curving ditch anomaly was found on excavation to be an earlier, Iron Age, feature and this continuity of landscape use is a common scenario seen within geophysical data. Closer examination of the results also reveals several fields cutting, or being cut, by a trackway and some following a slightly different axis to the main complex. While multiple phases can clearly be inferred from the geophysical data, it can be difficult to attempt to sort out even what appear to be simple stratigraphical relationships on the basis of the anomalies. In part this is a problem with all remotely sensed data; the techniques, both the data capture and processing, are designed to simplify the data into understandable images. The fact that buried features that are physically so close to one another that their signals may combine to produce a single anomaly illustrates the over-simplification that is inherent within the techniques and methodologies.

Villas

Although in Latin the word 'villa' means farm, in terms of the physical archaeological remains 'villa' is not so easy to define. Esmonde Cleary (1999) has suggested that it is 'a rural site exhibiting Roman-style buildings and architecture'. As a result the term is used to describe hundreds of sites ranging in date from the first to the early fifth centuries AD. Furthermore, they are seen to vary in size from a small dwelling to something on a par with a modern day mansion. As such, villa sites provide a source for some of the most impressive geophysical results that have been carried out, with examples to be found across the whole of the Roman Empire.

Villas commonly survive in modern rural landscapes and, as a consequence, tend to be free from later clutter that often obscures the results in more urban locations. The remains often comprise low foundations below thin topsoil and the walls follow fairly regular plans. As a result they are often revealed as crop or parch marks on the ground surface, although they are easier to plan from the air. The variation in the local moisture regime that creates the marks on the ground surface also generates anomalies within resistance data captured in area survey. However, when the aerial evidence is at a peak the resistance contrasts are generally past the optimum; in fact, the best time for a resistance survey over a villa, or any other stone structure, is during periods when the soils are moist in comparison to the building material. Many of the grander villa buildings incorporated hypocaust and areas of burning associated with fire boxes that fed the heating system. These factors, coupled with large quantities of brick, tile and fired clay that are often present, also make the sites attractive for magnetic survey. Because of the high level of activity in and around the villas, the associated field systems often produce clear responses, while the building itself tends to result in a distinctive area of magnetic noise. On several sites this magnetic 'noise' has been used to pinpoint where resistance survey, a slower technique, should be targeted. Although wall lines occasionally show in the magnetic data from villa sites, the clarity is seldom on a par with the resistance results.

The first example is from a site in Hampshire that was originally earmarked as having archaeological potential in the 1950s and 1960s, when ploughing brought to the surface a heavy scatter of Roman tile fragments (tegulae, imbrices and flue tiles) and red brick tesserae. In addition, in the northern face of a quarry adjacent to the site, wall foundations and floor surfaces, possibly associated with a small building, are still visible. It was decided to scan the area of the artefact scatters with a fluxgate gradiometer and this quickly identified a zone of magnetic noise that coincided with a slight earthwork platform. A detailed fluxgate gradiometer survey was targeted on this spot and confirmed the findings of the scanning. The erratic nature of the anomalies is due to the fired bricks, tiles and rubble associated with the former building. It is even possible to see one or two of the walls as a negative anomaly. However, it is only when the follow-up resistance results are viewed that a striking clear plan of a multi-roomed building becomes visible (**77**).

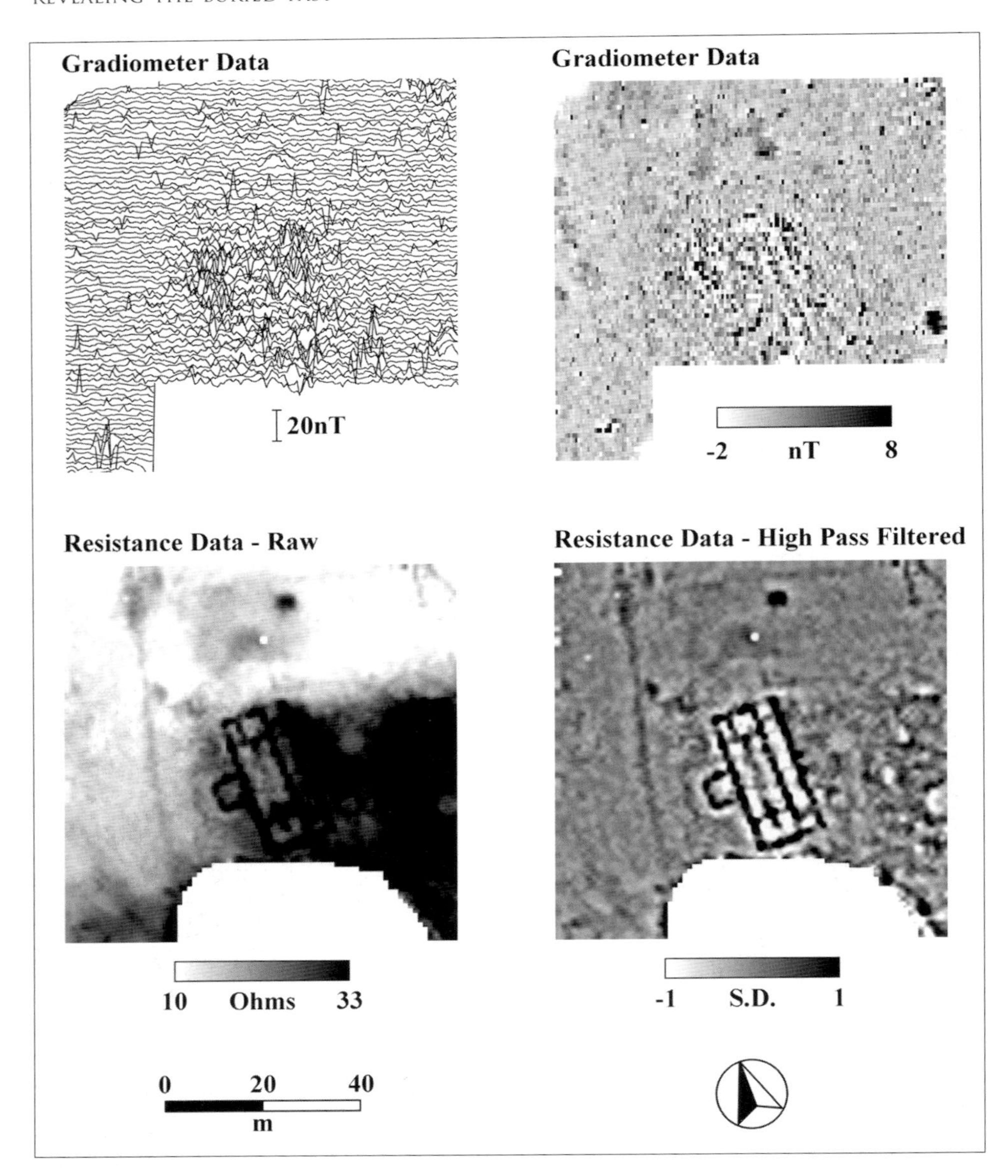

77 *Wanborough Roman Villa, Hampshire. Fluxgate gradiometer data (1 x 0.5m) and Twin-Probe resistance data (1 x 1m). The magnetic data indicate an area of noise that coincides with the villa building whilst the resistance data provide a clear plan of the building foundations.* © Crown copyright NMR

There appear to be nine or ten rooms or corridors and an apsidal room within an overall structure measuring approximately 38 x 18m. It is interesting to note that the building has no walls extending to the south, towards the edge of the quarry, indicating that the remains in the cliff section do not join with the main building. Other features in the data include what are presumed to be paths or boundary walls, plus some substantial pits, visible in both data sets.

A survey of the villa at Tockenham, Wiltshire revealed a wealth of both magnetic and resistance anomalies associated with the villa complex. The magnetic data indicate a series of small fields, enclosures, paddocks and/or garden features surrounding the main building. There are also several well-defined discrete anomalies which are indicative of large pits, perhaps used for storage or rubbish. Once again there is a distinct area of increased magnetic noise that corresponds with the foundations of the building. A resistance survey was targeted over the magnetic noise and the resultant plan includes detail of several rooms and corridors, plus one clear apsidal room which is believed to be a dining room or *triclinium* (**78**). There is a series of smaller rooms arranged around a small, rectangular intra-mural courtyard; on the northern edge is an unusual response that on partial excavation was believed to be an elaborate octagonal entrance. The range

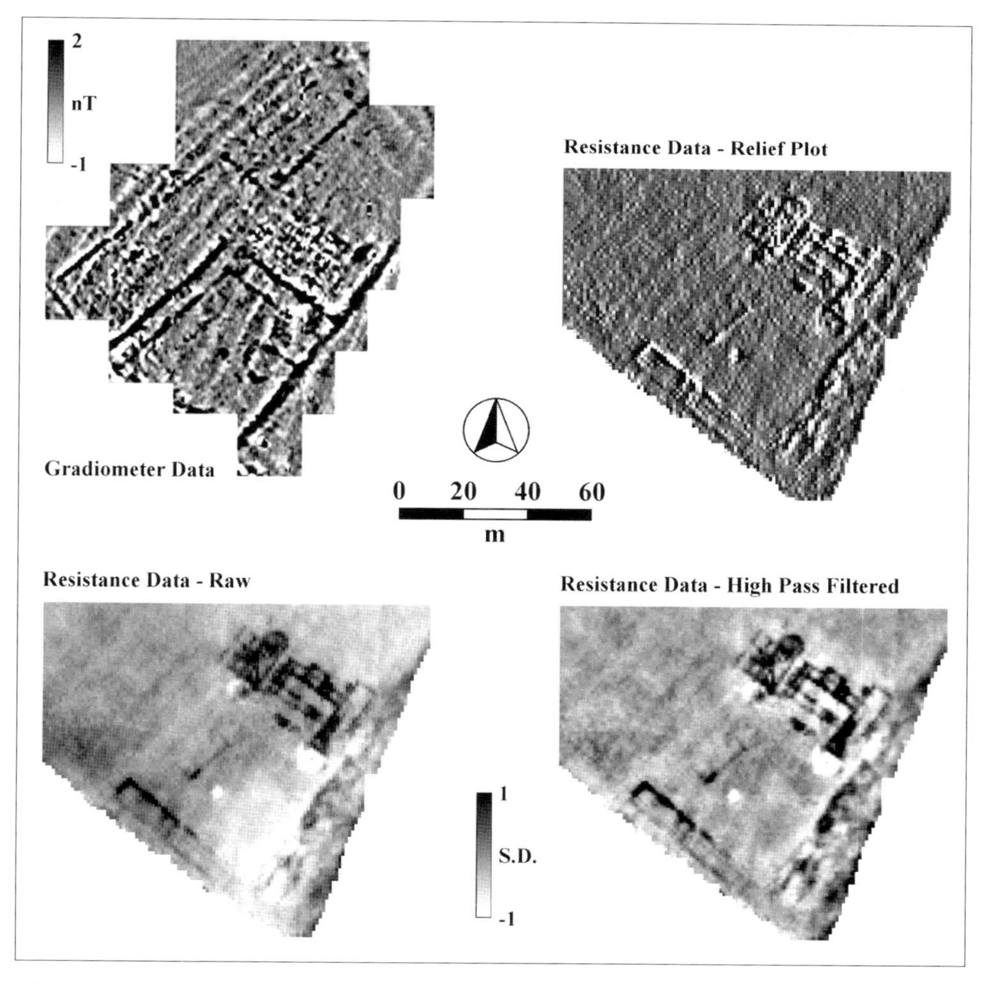

78 *Tockenham Roman Villa, Wiltshire. Fluxgate gradiometer data (1 x 0.5m) and Twin-Probe resistance data (1 x 1m). The two complementary datasets indicate differing elements of the villa complex*

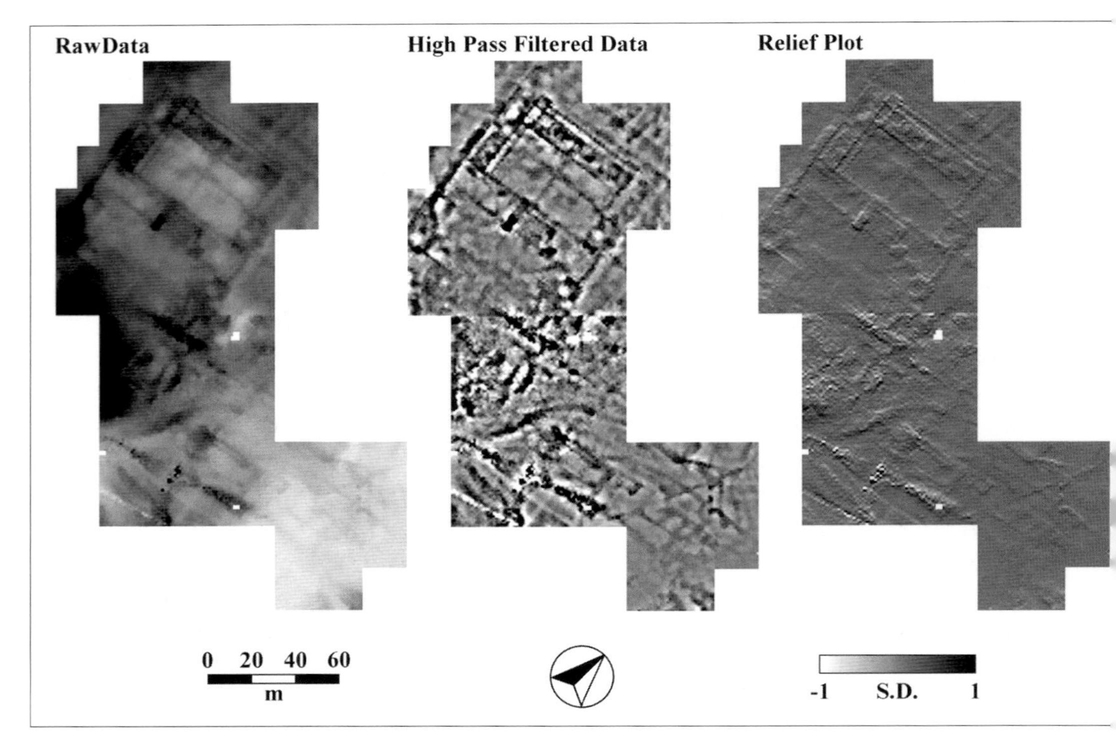

79 *Turkdean Roman Villa, Gloucestershire. Twin-Probe resistance data. 1 x 1m. A remarkably clear image of the villa complex was obtained during two visits to the site, both made in the spring but one year apart. It has proved impossible to seamlessly match the top half of the data from the first visit with the bottom half from the later survey*

of buildings attached to the eastern part of the courtyard is believed to be in the location of the bathhouse (Harding and Lewis 1997). The data also suggest more than one phase of activity, with wall lines on different alignments.

Moving up the scale in size, the villa at Turkdean, in the Cotswolds, afforded some of the best results we have ever obtained. The excitement of revealing for the first time ever a complete plan of the villa, was heightened by the fact the survey was undertaken for a 'live' broadcast for *Time Team* and watched by some 3 million viewers. Part of the resistance data are reproduced here, both 'raw' and following mathematical filtering to remove the effects of the geological background (**79**). The resistance survey covered about 2.4ha of land, with the core of the site covered in 1997 and extended in 1998. There are clear differences in the sharpness of the responses between the two periods of work. In the 1998 data the anomalies are considerably broader and this change relates to the weather conditions prior to each survey; although both data sets were collected in the spring, the '98 survey was later and consequently missed the peak signal resolution. As a result of this reduction of signal the data sets proved very difficult to merge. To produce the final versions of each display has involved an enormous amount of post-survey manipulation. Thankfully the hard work has paid off and it is possible to see very fine archaeological detail, from the existence of courtyards surrounding aisled corridors down to the level of individual doorways into some of the rooms. The plan provided a perfect template for deciding where to place the trial excavation trenches, and these confirmed the presence of wall foundations and floors, surviving immediately below the turf. The resistance data was complemented by over 6ha of magnetic survey (see Holbrook *et al.* forthcoming).

At Dinnington, the plan of a similarly large villa building (*c.*120 x 70m) was recovered, again using resistance as the primary technique, though in this instance it was possible to infer differing building phases within the data (**80**). The results suggest that an original building developed into the north corridor of a courtyard villa, by expanding to the west and east with the addition of two wings. The results also indicate the possible presence of a formal gateway structure. Excavation confirmed the existence of an earlier building and further identified a number of mosaics on the western wing of the building. These accounted for the broad areas of high resistance visible in the results. When the magnetic results are analysed, they indicate not only a distinctive area of noise that corresponds with the building but also clear trends in the data that reflect the modern-day ploughing. Excavation

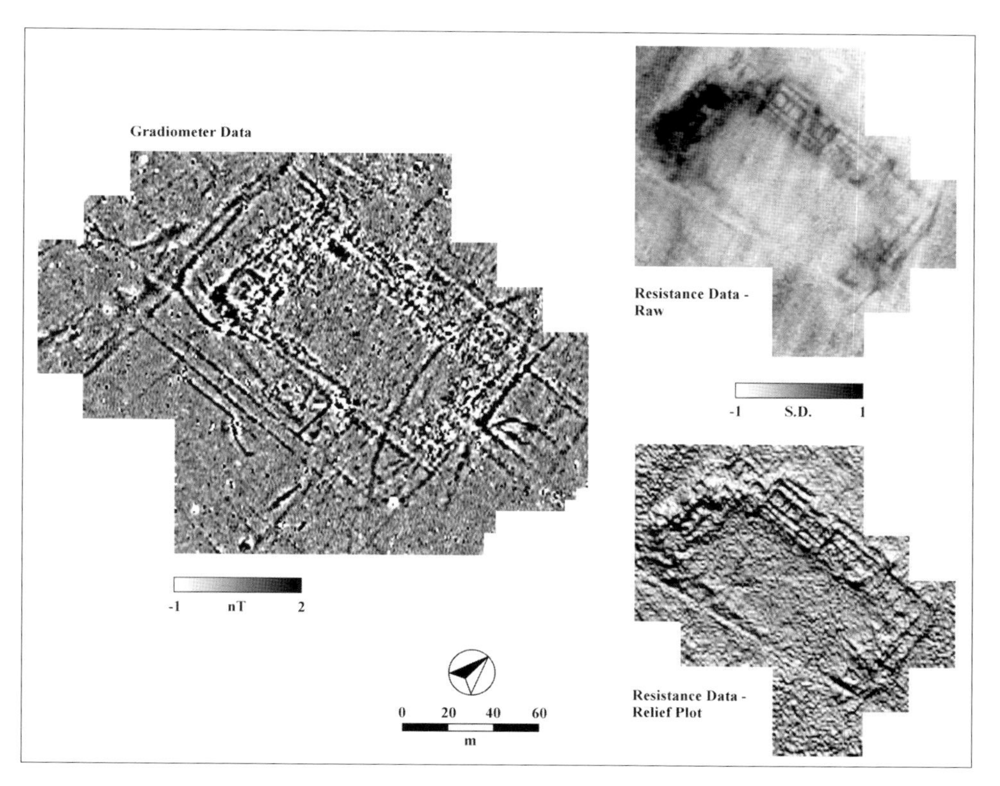

80 *Dinnington Roman Villa, Somerset. Fluxgate gradiometer data (1 x 0.25m) and Twin Probe resistance data (1 x 1m). The latter covers only part of the magnetic survey but is at the same scale*

of the mosaics showed dramatically the effect of this ploughing on the buried archaeology, with severe score lines cutting through the mosaics which survive within 15cm of the ground surface (**colour plate 18**).

Settlements and towns

Roman roads and particularly the junctions between them often became the focus for settlements. These developed in a variety of ways and the buried remains have long been the focus of aerial photographers. At Sedgefield, County Durham, we were able to map an unusual example of one such settlement. Although prior to the geophysics tentative enclosures had been noted on aerial photographs and numerous metal detector finds had been recovered, it was not until the gradiometer survey was carried out that the complexity of the site was appreciated. Straddling a trackway or road is a regular pattern of small enclosures/paddocks that form a pattern known as a 'ladder' settlement. Whilst such settlements are relatively common in the Midlands and southern Britain, this was apparently the first time a complex had been surveyed so far north. Small-scale excavation found a distinct lack of actual occupation deposits amongst the finds. It would appear that the southern half of the site comprised workshops, stock enclosures and small-scale industrial type activity, in the form of kilns and metal working areas. One clear magnetic anomaly proved on excavation to be a small pottery kiln that had survived intact (**colour plate 19**). A scatter of larger anomalies, visible in the magnetic data, was interpreted on excavation to be clay pits that had become back-filled with a mixture of magnetically-enhanced deposits. While time did not permit the full mapping of the extent of the settlement at Sedgefield, the English Heritage survey at Owmby, Lincolnshire provides a largely complete picture of ribbon development along Ermine Street (**81**). The survey is a good example of the link between enhanced magnetic susceptibility in the topsoil and the core of the archaeological site.

Survey during the early 1990s at the Roman town of *Calleva Atrebatum* (Silchester) reveals the value of differing data processing. There are clearly important elements of the town within each data set, but the problem is the scale of the site. The surveys shown in figure **82** only cover a small percentage of the area within the town walls of Silchester, some 40ha in total. Without recourse to the extensive aerial photographic or excavation evidence, it is impossible to understand how representative the geophysical results are of the interior of the town, or to comprehend the significance of the detailed response. A larger, more recent survey of Silchester, by English Heritage, has shown that magnetic data correlates well with the known layout of the town (Martin 2000).

Moving to the top of the scale in size are the results from the Roman city of *Viroconium* (Wroxeter). Wroxeter covers about 78ha of land near to the Shropshire town of Shrewsbury and was the fourth largest town in Roman Britain. The town is largely under grass and has been researched by antiquarians and archaeologists for many centuries. While the site has been intensively studied from the air (Wilson 1984), modern excavation, primarily by Philip Barker, has looked at less than one per cent of the town (White and Barker 1998). As a result the interpretation of the site had to come from the aerial evidence, which suggested that perhaps 40 per cent of the town was blank. A research design was prepared that integrated a magnetic survey of the whole of the town, thereby avoiding expensive and damaging excavation (Buteux *et al.* 2000). Earlier work in the environs of Wroxeter had often produced poor responses due to the local soils which are seasonally waterlogged, high in clay content and occasionally fail to produce magnetic anomalies above the soil background. While that may not have been a good omen for undertaking a large-scale survey in this area, the fact that the survey was within a Roman city improved the odds of success. In fact the soils inside most towns or cities, whether ancient or modern, are completely altered from their parent formations due to the immense amount of work that has been undertaken on them; they are often black with organic and burnt material.

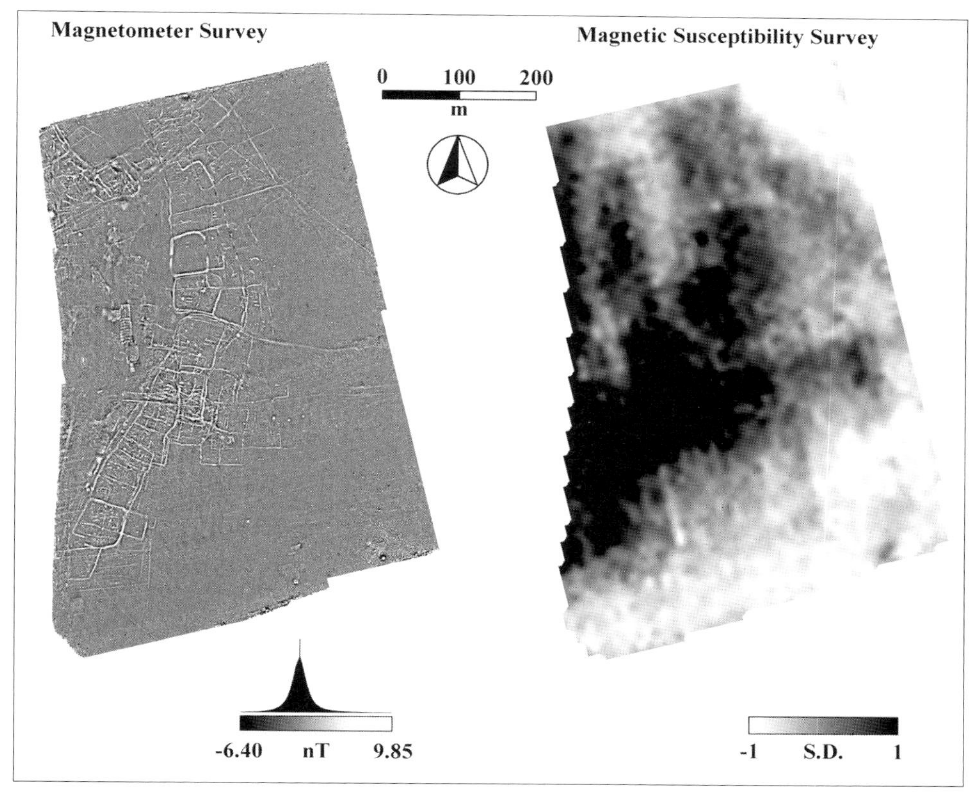

81 *Ownby, Lincolnshire. Fluxgate gradiometer data. 1 x 0.25m. Survey and image courtesy of AM Lab, English Heritage. Note that the plot has higher magnetic anomalies plotted in white. Magnetic Susceptibilty data 5 x 5m grid*

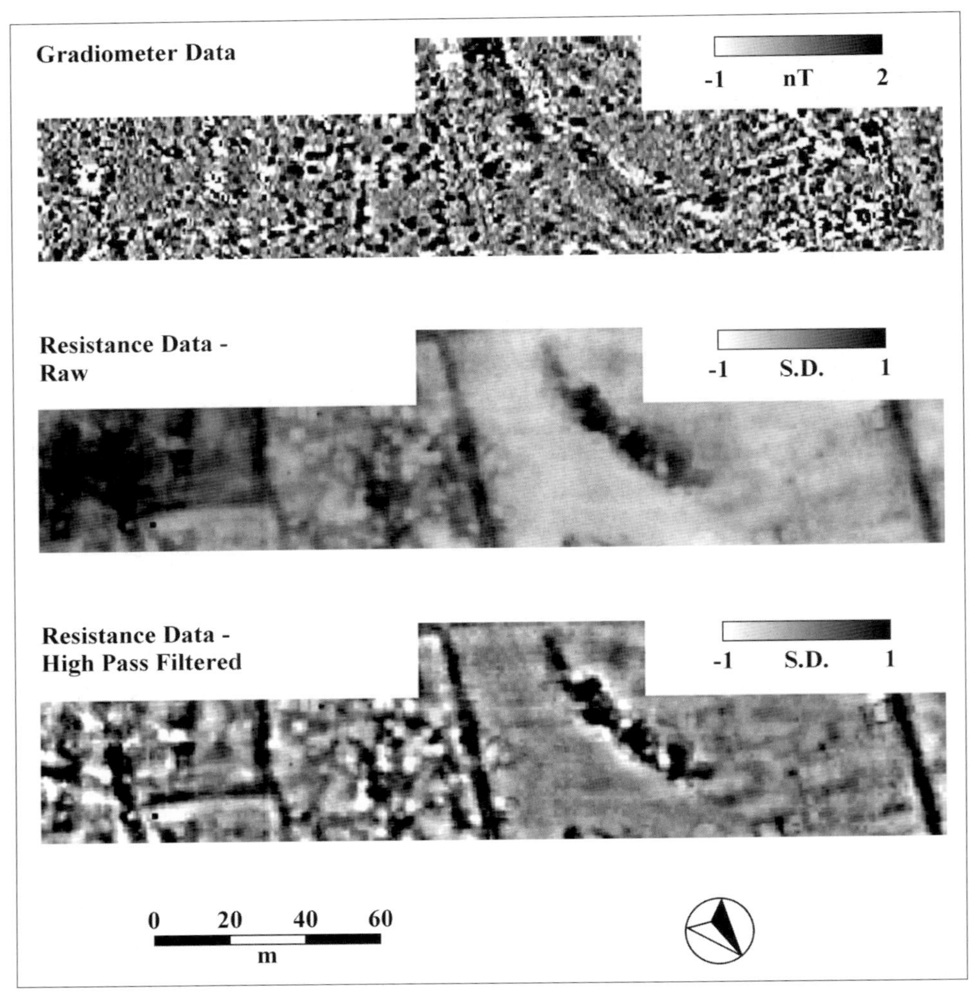

82 *Silchester Roman town. Fluxgate gradiometer data (1 x 0.5m) and resistance data (1 x 1m).* GSB survey for Professor Mike Fulford, Reading University

The detail from one of the 15 fields that were surveyed can be seen in figure **83**. The wealth of information in this image is typical of the density and complexity of archaeology detected within ancient towns using a fluxgate gradiometer. The data are split into two by responses from the later Roman defences and are clearly different either side of them. To the south can be seen part of a regular road grid as well as negative responses from stone-built buildings that are aligned with the road system. To the north of the defences the anomalies become more randomised, with a significant number of responses attributed to pits as well as a few suggesting industrial origins. There are also numerous ditch anomalies as well as areas of disturbance that indicate increased soil noise due to disturbed archaeological features or strata.

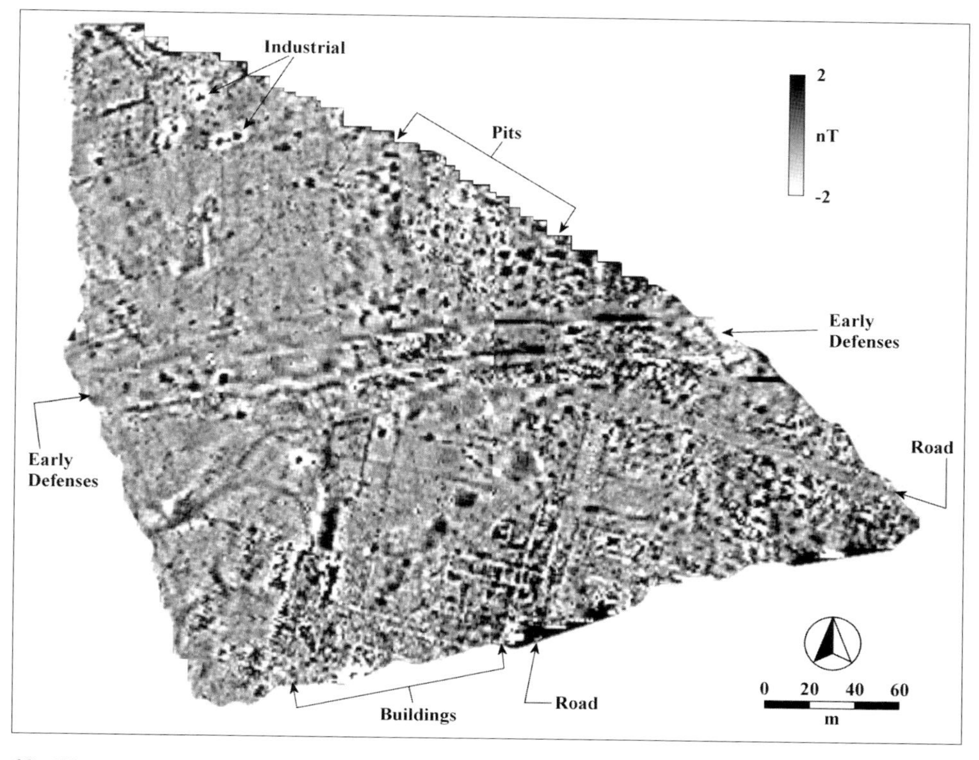

83 *Wroxeter Roman city. Fluxgate gradiometer survey (1 x 0.5m). Part of the complex indicating defences, roads, buildings, industrial activity and pits.* GSB survey for BUFAU and Wroxeter Hinterland Project

The interpretation of the results had to be undertaken in a highly systematic manner as the overall data set contained nearly 3 million readings, covered over 70ha of ground and was undertaken by two different groups (GSB Prospection and the AM Lab, English Heritage). A further complication is that from the first block of data collection to the final interpretation the project covered a four-year span. Prior to the main publication in *Archaeological Prospection* a classification of the anomalies was agreed (Gaffney *et al.* 2000). This is reproduced in **Table 2** and, while it is specific to the Wroxeter site, it is useful to understand how the collaborators broke down the components of a complex magnetic data set.

The overall magnetic data set from Wroxeter is reproduced in figure **84** and the city can be seen to be virtually filled with anomalies of archaeological interest. In fact the magnetic survey shows a remarkably well-developed city with some rare insights into economic, social and functional zonation. However, the geophysical data can only be pushed so far; there is little control over the temporal variation and it is assumed that the settlement and activity evident in the data shows a maximum extent within the town's defences. No matter how sharp the image is it should not be supposed that we know all there is to know about Wroxeter.

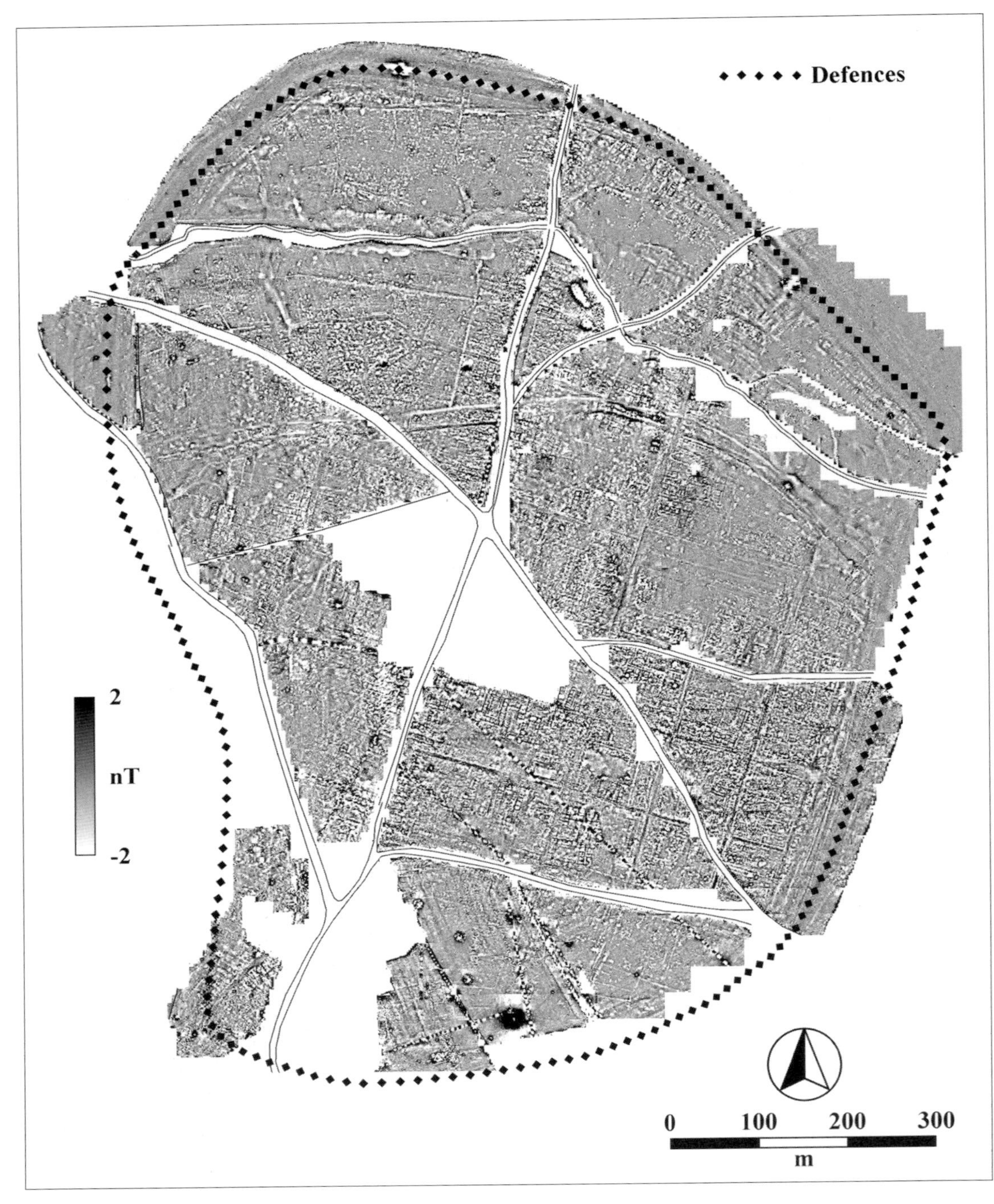

84 *Wroxeter Roman city. Fluxgate gradiometer survey (1 x 0.5m). Approximately 70ha of data, more than 3 million data readings, were collected by GSB and the AM Lab during separate visits over a 4-year time span.* Work carried out in conjunction with BUFAU as part of the Wroxeter Hinterland Project

Fortified sites

There were a wide range of military installations used by the Romans; these included 'temporary/marching' camps through to legionary fortresses, and they varied in size and complexity from less than one hectare to several tens of hectares. Close military control of certain areas was maintained through a series of forts and fortlets served by an intricate road network (Hanson 1999). In Britain, at the hubs of the network were the legionary fortresses of York, Carleon and Chester; one level down were the main forts and then fortlets; watch-towers/signal stations, or turrets as they are called on Hadrian's Wall, formed the smallest bases. Some of the installations were simply ditched and banked enclosures with temporary encampments, while others were occupied longer term and had stone buildings and defences. From a geophysicist's point of view, magnetic and resistance surveys are usually most productive, though the increasing use of GPR on the continent demonstrates the extra level of information that can be obtained if time permits.

As we have already shown, some of the earliest geophysical work in Britain was carried out in the 1950s, in advance of widening the A1 trunk road. It is

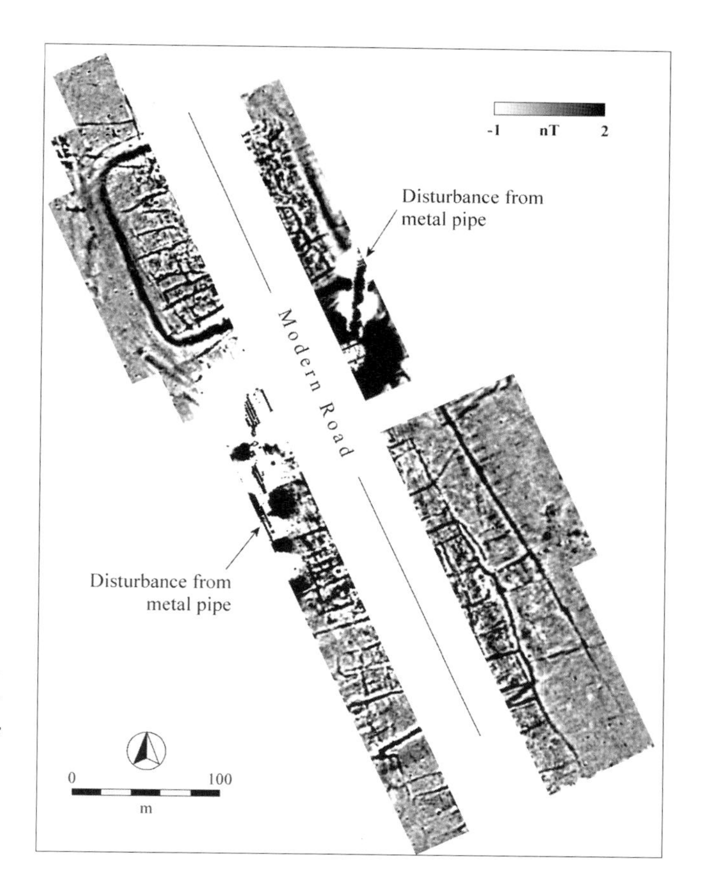

85 *Healam Bridge, Yorkshire. Fluxgate gradiometer survey (1 x 0.5m) carried out in advance of proposed widening of the A1 trunk road. The results indicate a Roman fort and ladder settlement which both straddle the line of the road.* GSB survey for Ed Dennison/AWP/BHWB on behalf of the Highways Agency

somewhat ironic, therefore, that almost 50 years later we found ourselves carrying out similar work prior to the A1 being upgraded to motorway status. During this project geophysics was employed extensively to evaluate large stretches of land and several new sites of importance were located. Of particular interest was the mapping of a previously suspected fort and settlement at Healam Bridge (**85**). Here the existing A1 is seen to pass right through the centre of the site and a so-called ladder-type settlement that extends to the south. There is a dramatic difference between the levels of magnetic enhancement inside the fort compared to the immediate area outside. In places the fort ditches appear to overlie smaller ditches on a differing alignment, perhaps indicating that the fort was constructed over an existing strategic site. Given the dangers associated with stratigraphical relationships, it may be that we are reduced to saying that there is simply more than one phase of activity. Unfortunately the interface between the presumed military and civilian settlements has been obscured by a modern gas pipeline. The survey further shows a good example of continuity of road usage over a 2,000-year span, with the A1 following the exact line of Dere Street, one of the major Roman arterial roads.

Industrial sites

Pottery (and tile) kilns have long been considered 'classic' targets for magnetic survey. Having been fired to high temperatures, when they cool down the magnetic minerals in the clays realign with the Earth's magnetic field and retain a relatively strong field of their own. As we have seen in chapter 1, this thermoremanent magnetisation was first recognised in the 1950s as making kilns suitable targets for detection by proton magnetometers. Since that time, probably hundreds of kilns have been surveyed and numerous theoretical models produced, attempting to predict the magnetic field associated with the different types of structures.

In their simplest form a kiln will comprise a central chamber, where the pots or tiles are fired, set within the walls of the structure. Thus when a magnetometer passes over a kiln the instrument will see first one wall, then the interior and then the second wall. The resultant anomaly will comprise two peaks either side of an enhanced anomaly. The reason for this is that the highly fired clay walls have a stronger field than the interior which lacks the mass of structural clay. In the accompanying example (**86**) a survey over an expected Romano-British kiln complex in the New Forest produced a strong anomaly of over 100nT, which has a double peak in two of the traces. Clearly there is nothing in the geophysical data to suggest a Romano-British date; in this instance the supporting archaeological evidence in terms of pottery wasters provides an implied date.

Chesters Villa, situated on the northern shore of the Severn Estuary, has been subjected to a number of geophysical surveys. During a resistance survey over the main buildings in 1987, topsoil samples were collected from over the villa

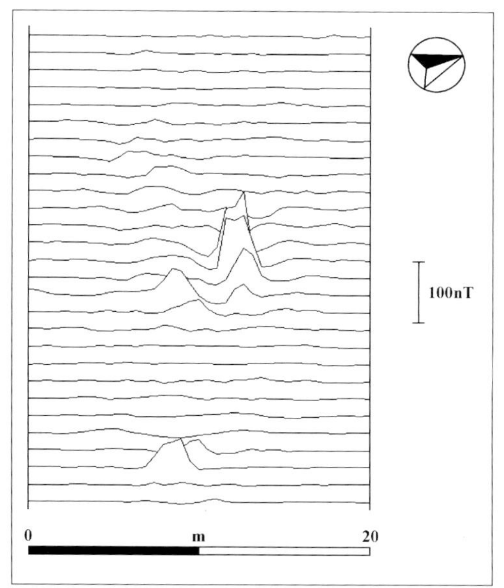

86 (Above) *Sloden enclosure, New Forest. Fluxgate gradiometer survey. 1 x 0.5m. Results demonstrate a 'classic' double peak over a suspected pottery kiln of Romano-British date.* GSB survey for RCHME. © Crown copyright NMR

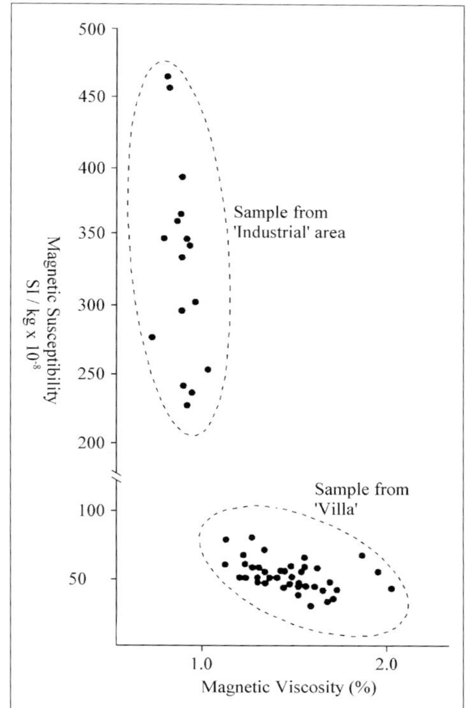

87 (Right) *Chester Roman Villa, Gloucestershire. The diagram shows a plot of the magnetic susceptibility vs magnetic viscosity, and illustrates the different signatures between soils derived from settlement and industrial areas. The soil samples were collected over a 20m grid that was established for a resistance survey. Prior to excavation a gradiometer survey revealed very strong responses which were related to industrial processes and ditches infilled with debris from that activity*

and a nearby spread of dark earth. The field was known to have produced substantial amounts of tap slag and charcoal and after ploughing it was clear that much of it was to be found over the spread of the dark earth (Fulford and Allen 1992). The soil samples were tested for magnetic susceptibility as well as magnetic viscosity as it was hoped that the two areas would produce different signatures. In fact the results classically illustrate the difference between soils that are derived from settlement (moderate magnetic susceptibility and viscosity) and those derived from industrial activity (high susceptibility and low viscosity) (**87**). This small-scale soil analysis definitively supported the fieldwalking evidence. Excavation at the site revealed masses of tap slag and evidence for a number of furnaces. A fluxgate gradiometer survey revealed the extent of the industrial site and it was found that it was largely contained within an enclosure (Gaffney and Gater 1992).

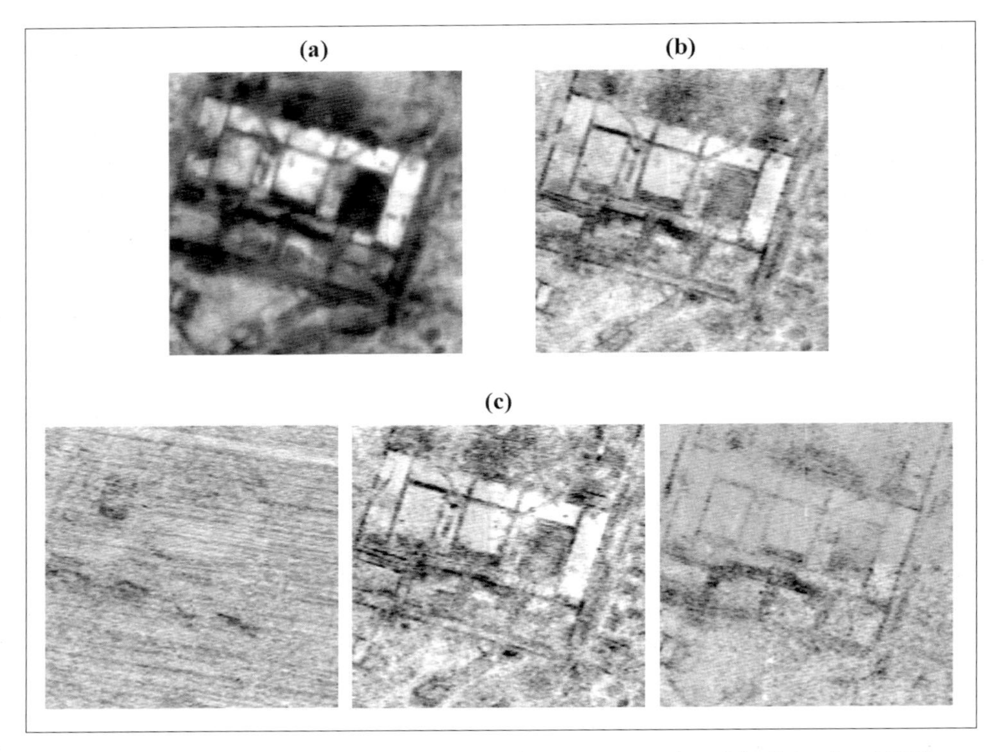

88 *Data collected at the Roman town of Carnuntum: (a) is a resistivity plot, while (b) is a broad GPR time slice covering a depth of about 1.5m of ground. The two images are highly compatible. However, the GPR data has added information in that narrow slices can also be produced to reveal the variation with depth. In (c) three images are shown with the shallowest on the left and the deepest time slice on the right. For a more extensive discussion of this survey see Nebauer* et al. *2002. The images have been supplied by ZAMG Archeo Prospection (R), Vienna*

Wider perspective

Geophysical surveying of early historic sites has a long history. In fact a great number of the 'classic' geophysical surveys have been undertaken on large sites such as those at Sybaris and Xanten in Europe. These sites have a common theme in that they are high status sites and have produced equally impressive results. While it is true that the majority of sites from this period were graced with more humble abodes, many of the recently published examples are still from the tradition of grand sites. Surveys from all over the Roman world have consistently delivered significant results; in fact from the heart to the extremities of the Empire measurements have been collected using virtually every geophysical technique and productive results are nearly always obtained.

The abundance of good quality, complementary data from many of these sites has led to an explosion of publications that base their interpretations on multi-method analysis. This work has ranged from lowly Romano-British villas (e.g.

Corney *et al.* 1994) through villas owned by individual Emperors (Piro *et al.* 2003) and incorporates strategies for the investigation of whole towns (Gaffney *et al.* 2000, Neubauer *et al.* 2002). Data from some of the work at the Roman town of Carnuntum in Austria can be seen in figure **88** and this high resolution, multi-technique approach has allowed the researchers to produce a very specific interpretation using the following categories: mortar, foundation, hypocaust, drain, wall, staircase and tiles (Neubauer *et al.* 2002).

The concept of high resolution, multi-technique studies has become fairly routine across much of the Mediterranean zone and encompasses research not only in the Roman but also the earlier part of the historic period. Again the scale of work is highly variable but increasingly the surveys are embedded within sophisticated research designs. For example the geophysical work on the island of Alonnisos (ancient Ikos) has focused on the production of amphorae during the fourth century BC, but is part of a wider archaeological, ethno-archaeological and geomorphological study to inform on the diachronic land use of the island. The overall objective is to understand how the economy supported a minor classical polis, which in turn will feed into the debate on how the island interacted with the mainland of Greece (Sarris *et al.* 2002).

Summary points

• During the historic period, the increasing use of stone as a building material is important in the choice of technique. In mild temperate climates, such as those found in Britain, the technique that is most often used on sites that are suspected to have this type of buried remains is resistance survey. In countries where the soil is naturally dry and a good electrical contact is difficult to achieve then GPR systems come to the fore.

• Magnetic surveying remains important in prospecting for sites from this period because the settlement sites were often large, long-lived and had a significant element of intensive burnt features, associated for example with hearths or industrial processes. The settlements often produce an imprint of enhanced magnetic susceptibility on the surface of the soil that can be mapped using magnetic susceptibility.

• As the sites are often complex or multi-phase, it is clear that a battery of techniques can be usefully applied. This, along with appropriate analysis, may result in closer integration with other information sources and allow a more coherent archaeological answer.

8

SURVEYING ON LATER HISTORIC AND MODERN SITES

The time span covered in this chapter is roughly 1,500 years in length and embraces a very wide range of archaeological features; and the examples chosen aim to represent the variety of geophysical responses that can be obtained. We start with churches and monastic sites, move on to historic houses and castles and gardens, touch upon agricultural features and industrial features and end with a look at military sites. The examples chosen include a number with anomalies that appear to have an archaeological origin but are in fact very recent phenomena. It will be clear from the list of type sites that there are many that we have left out; some type sites have not been included because they do not produce many anomalies of any great clarity. In particular, settlement sites in the immediately post-Roman period often produce geophysical signals that are far from characteristic. Where measurable signals are produced they are rarely characteristic and can easily be confused with anomalies from either geology or prehistoric remains. There are of course some exceptions, most notably Lyall and Powlesland (1996) that have been able to disentangle elements of the early/middle Anglo-Saxon from the Late Roman. This has largely been achieved by re-surveying at high sample density (0.25 x 0.25m) after the topsoil has been removed. Other researchers have incorporated geophysics into a broader scheme of works that includes the analysis of trace element (chemical) variation within the topsoil (Aston *et al.* 1998). It is likely that painstaking documentary and environmental research is required for finding 'Dark Age' sites and once found, high sample densities may be required to map them in detail. As a result, the methodologies for prospecting for Dark Age sites are still open to debate and as a consequence we will not pursue them here. In general the key to the success of prospecting techniques on later historic sites is, in many cases, the use of stone and brick as construction materials as well as an abundance of other readily detectable material.

Churches and monastic sites

The earliest Christian sites appear in Britain in late Roman times, but the number of recorded geophysical surveys on ecclesiastical remains from this period is small.

It is not until stone churches appear in the landscape that geophysics starts to play a role in mapping such structures. This is not to dismiss the use of techniques on early church sites; it is simply that the results are rarely of visual merit.

In the late 1980s a survey was undertaken at Spon in Wales, the site of a presumed single-celled chapel. It was thought, on the basis of documentary and place-name evidence, that the structure lay within the confines of a field threatened by an extension to an existing opencast mine. The brief was to locate the buried remains and then define their extent, in order that they could be fully investigated prior to a proposed development. Unfortunately, there is no easy method of 'scanning' when it comes to resistance survey and unlike with Roman villas, where magnetic survey can pinpoint areas of characteristic noise associated with the buildings, there tends to be no such detectable enhancement associated with church sites. A good approach to this problem would be to undertake some systematic sample across the site. An alternative approach is to carry out a systematic detailed survey on a grid by grid basis until anomalies of potential are identified and then the grid can be extended to encompass the site. At Spon the latter was chosen as there were a number of topographic clues to indicate the possible position of the church; in this case it imposed a strict sample scheme that would have taken too long and would have potentially missed the small rectangular building. Given that we were looking for a chapel it was decided to orientate the grid on a south-west/north-east axis in an attempt to traverse any surviving walls at a 45 degree angle. The data were downloaded after each survey grid and the results analysed whilst still in the field. Despite a varying background due to the underlying geology, the anomalies associated with the buried church walls were very distinct (**89**); there was a clear rectangular anomaly measuring

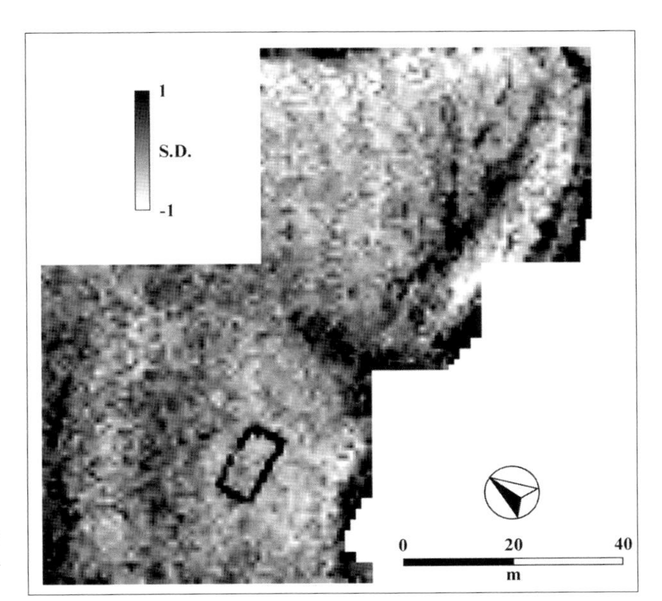

89 *Twin-probe resistance data showing a single-celled chapel. Original data collected on a 1 x 1m grid.* GSB survey for the Clywd-Powys Archaeological Trust

some 12m long and 8m wide and orientated east-west. Fortunately the remains have been saved as the planning application was subsequently turned down.

A rather more substantial church was located at the site of Athelney Abbey, in Somerset. This was a particularly satisfying and exciting survey for us, as the work was carried out as part of the first *Time Team* programme ever transmitted. The clarity of the results provided a remarkably clear picture of the abbey church and the surrounding monastic buildings (**colour plate 23**). The results were particularly surprising because the site had originally been surveyed some six years earlier and the results proved negative. The main reason for this is that the initial resistance survey was carried out when there was little or no moisture contrast in the soil. We were lucky, therefore, in that when we carried out our survey, in the spring, ground conditions were ideal for survey, as there was optimum moisture contrast between the wall foundations and the surroundings. The survey demonstrates how important the gross seasonal variation in the weather can be when carrying out resistance surveys, as the signal to noise ratio can vary immensely.

It is often the case that the church building survives better than the associated remains and geophysical survey is used to fill in the gaps of the monastic plan. One particularly successful collaborative investigation we were involved in was with Professor Mick Aston and the Royal Commission on Historic Monuments of England (RCHME). Using mainly twin probe resistance survey we examined nearly all of the monastic complexes in Britain belonging to the Carthusian order. This provided a unique opportunity to map the remains and produced results on a par with those recorded on some of the best Roman villa sites. Given that the monastic sites often comprise shallow stone foundations then this success should not be surprising. The ecclesiastical complexes also tend to follow similar plans though the detail varies from site to site. One distinctive feature of the Carthusian houses is the presence of individual cells, attached to the cloisters, where each monk occupied his own two-storey house and garden. Unlike other monastic orders, the Carthusians did not live a communal life, so each cell was in effect a personal monastery in miniature in which he lived and worked (Coppack 1990; Coppack and Aston 2002). One of the best-preserved sites is at Mount Grace, Yorkshire, where the standing remains provide a good picture of the original layout and detail. Due to this excellent preservation, no geophysics has been required, though the plan provides a working template for the interpretation of the resistance results from other sites, such as Axholme and Hinton Charterhouse.

At Axholme, the priory was founded in 1396 by the Earl of Nottingham and dissolved in 1539. In total contrast to Mount Grace, there are no upstanding remains, apart from a cellar beneath a post-medieval farmhouse that stands on the site. Although there are numerous extant earthworks it is thought that the majority are associated with post-dissolution houses and relict garden features, some of which reflect underlying monastic alignments. The clarity of the resistance survey plan is therefore quite remarkable. The results are presented as a mixture of greyscale and colour-scale relief plots, after mathematical filtering has been carried out to remove the geological background (**colour plate 20**). The plots present a virtual 3D effect that gives the impression of a terrain model. Individual cells within the cloister range have been identified, together with further buildings to the north. The presumed western arm of a moat is also apparent in the data, the southern and eastern arms are visible on the ground as water features, and a building in the north-western corner of the survey is interpreted as a gatehouse with a track leading into the complex. Perhaps the only disappointing aspect of the survey is the lack of a well-defined church due to the presence of modern buildings and yards.

Hinton Charterhouse, to the south of the city of Bath, was founded some 150 years earlier than Axholme. It was the second Carthusian house in England and spanned the years 1232 until 1539. Elements of the original buildings survive as standing structures and this evidence, together with piecemeal excavations, has helped to provide a layout based on the great cloister. When the earthworks survey and the geophysics were completed the monks' cells were clearly recognised, although elsewhere the complexity of the results made archaeological interpretation difficult (**90**). The reason why monastic sites produce good results is that stone is used widely in their construction and this provides a measurable contrast compared to lowly habitation sites from this period that are made of earthen banks.

An unusual site that we recently investigated was a former leper hospital outside the city of Winchester. Dating from the twelfth century AD, today nothing survives above ground level, but there is good documentary evidence indicating the location of the complex. This was fortunate because the geophysics was carried out as part of a *Time Team* shoot and we only had three days to complete the survey. Armed with early maps we were able to narrow down the area that had to be surveyed and almost immediately we succeeded in pinpointing part of the almshouses. Initially only one 20m grid of resistance was surveyed (prior to this we were using magnetics to try to locate the boundary of the hospital) and the results led us to hope that we would obtain a clear plan of the remainder of the buildings on the following two days. Unfortunately we had not predicted the dramatic change in weather that was forthcoming. On the second and third days there was torrential rain and puddles of water rapidly built up on the ground surface. What had previously been dry ground suddenly became totally sodden and as a consequence most of the electric current passed along the ground surface and little depth penetration was achieved. In addition a high density of brick and tile became visible on the surface of the field and this made electrical contact very difficult and at times impossible. Even re-surveying the grids with slower data capture times or logging the data manually failed to overcome the high percentage of bad data points (**91**). We were lucky in that the main elements of the hospital complex were identifiable within the results: the church lying to the south of the almshouses and the master's lodge linking the two to form a sort of cloistral arrangement. The linear anomaly to the east and north is the medieval boundary.

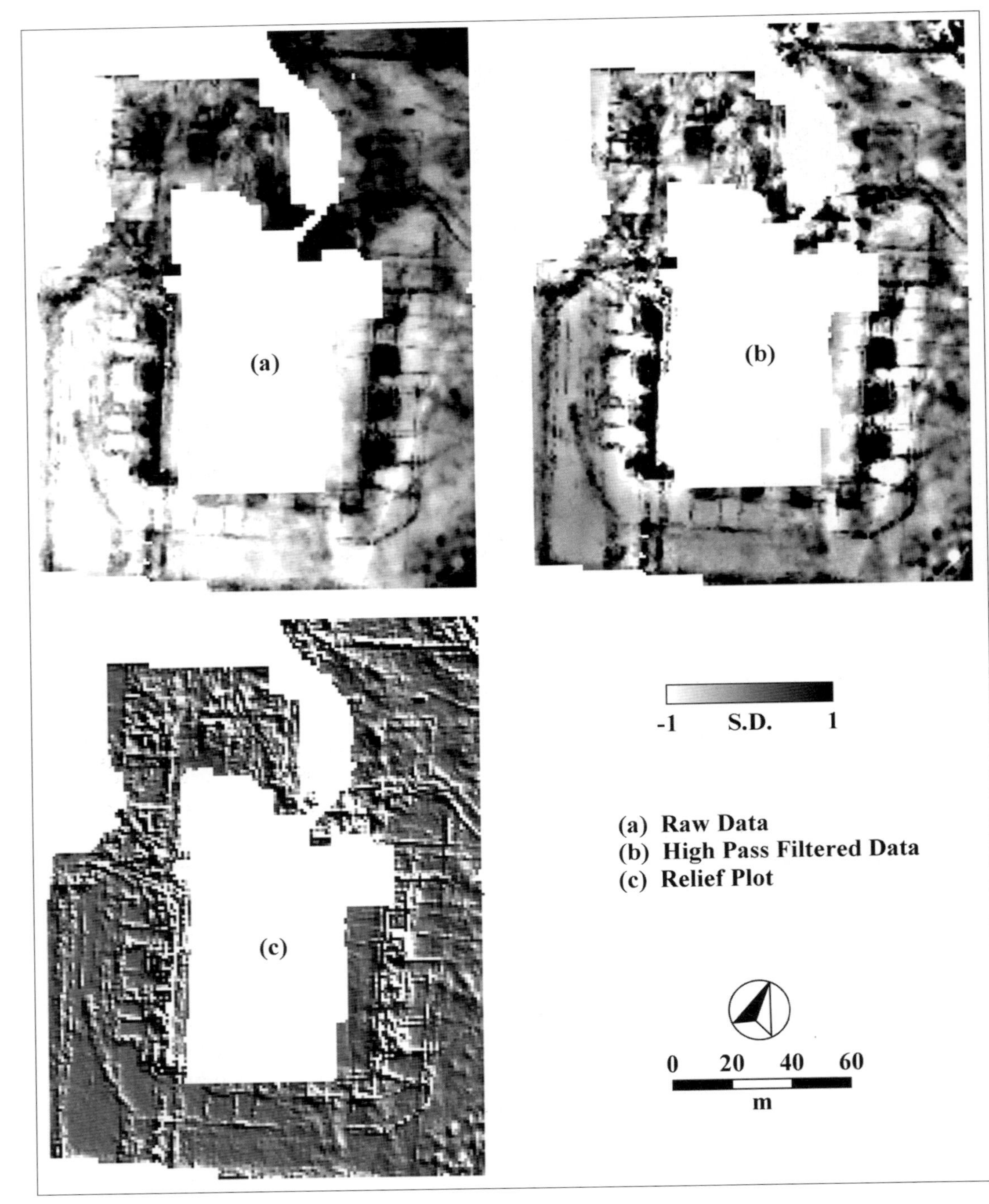

90 *Hinton Charterhouse Carthusian Remains. Twin-probe resistance data at 1 x 1m intervals.* GSB survey for RCHME. © Crown copyright NMR

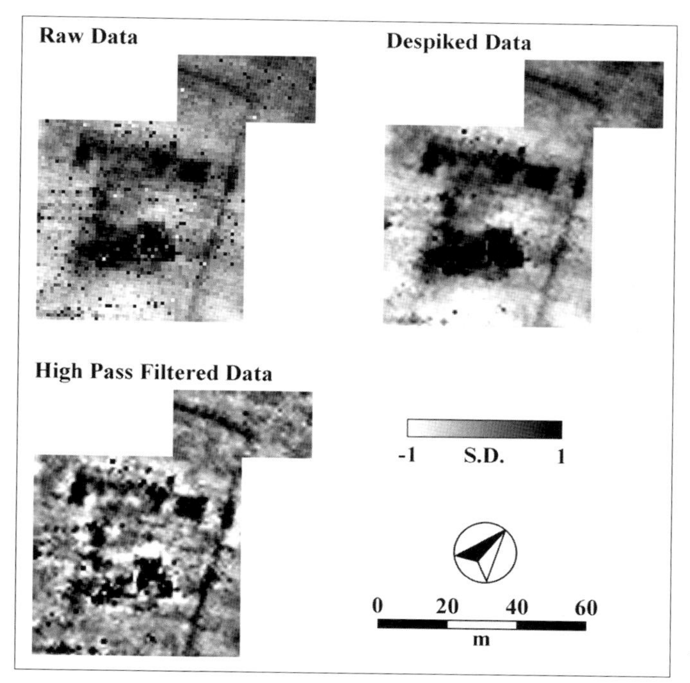

91 *St Mary Magdalene leper Hospital, Winchester. Twin-probe resistance data at 1 x 1m intervals. The results highlight difficulties in collecting data when ground and weather conditions are poor*

Historic houses and castles

The archaeological significance of results from geophysical surveys on medieval villages are variable; the buildings can be timber built or perhaps clay lump, and while rubbish pits, hearths, ovens and other fired features are readily detected using magnetic survey, only a very partial picture will be recovered. There are many instances of surveys on upstanding earthwork sites, particularly deserted or shrunken medieval villages, where the results are poor. This is due to the lack of measurable contrasts and the fact that topographic effects can dominate the data. This tends to mask the buried archaeology and the geophysical results can often be seen to reflect the topographical variation.

By way of contrast, the prestigious sites associated with the ruling elite, such as castles and stately homes, might be thought of as ideal targets for geophysics. Although the nature of the construction materials should be conducive to geophysical investigation, the fact that they often survive as partially standing structures leads to complications; the areas available for investigation can be quite small and the consequences of landscaping or consolidation of the grounds cannot be underestimated.

At Stafford Castle the survey report, which covers an extensive programme of geophysical investigations, summarises the difficulties that castle sites can present when carrying out geophysical survey:

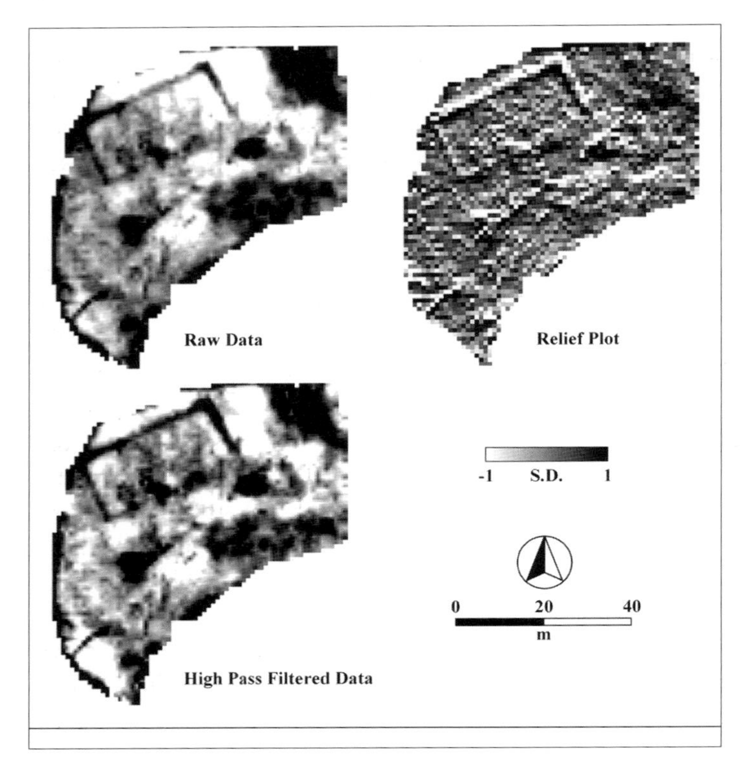

92 *White Castle, Wales. 1 x 1m Twin-probe resistance data indicating a previously unknown rectangular building.* GSB survey for CADW

> dumps of demolition rubble, in places 1m deep, are known to mask areas containing the subtle evidence for timber structures . . . the areas are neither flat (the slopes . . . are often over 30 degrees) nor devoid of extant or known stone structures . . . hardly ideal conditions for the location of structures . . . moisture variations produced by the undulating nature of the ground surface produced anomalies that have obscured responses from archaeological features . . . earthworks produced similar responses . . .
>
> (Darlington and Shiel 2001, p.146)

Yet despite these difficulties, several features of archaeological interest were identified at Stafford and we have worked on numerous other surviving castle sites where carefully targeted geophysics has produced good results. One such example is at White Castle, in Wales, where a rectangular building, a probable aisled structure measuring *c*.35 x 20m, was discovered inside the standing castle walls (**92**). An example that illustrates the type of response that can be obtained when a castle has been largely demolished comes from Kasteel Schonauwen, in the province of Utrecht in the Netherlands. The castle was built in the early fourteenth century, although it was expanded significantly during the seventeenth and eighteenth centuries. The area was surveyed by RAAP Archeologisch

Adviesbureau with a Geoscan RM15 earth resistance meter on a 1 x 1 m grid with a probe separation of 1m. The choice of a wide probe separation was due to the depth of overburden; the matrix consists of fluvial clay and, in the south-west, of peat. The castle was built with brick and the contrast between this and the surrounding clay matrix is dramatic (**93**).

The same difficulties encountered on castle sites also apply to historic house sites, but one of the main difficulties that often occur is being able to interpret the sheer complexity of geophysical responses that are recorded. This is often a result of occupation and rebuilding over a long period of time, which frequently involved contraction and expansion of the complex, something not so apparent on castle sites.

For example, at Rycote Manor in Northamptonshire, a plethora of both resistance (**94**) and magnetic anomalies were mapped. The interpretation was complicated by trees and shrubs, upstanding earthworks and the depth of archaeological stratigraphy. In places, excavations showed that some walls had been robbed out and backfilled with soil, while elsewhere they were backfilled with stone and brick rubble, resulting in differing responses. Also, some of the walls were built of brick, others of stone, which resulted in differing magnetic responses. Former fireplaces and oven bases produced strong magnetic anomalies but it was not always possible to differentiate between these and demolition spreads of bricks, or rubbish pits. In such instances, while predictions can be made, it is only with the benefit of excavated evidence that the data can be re-analysed and then re-interpreted with any degree of confidence.

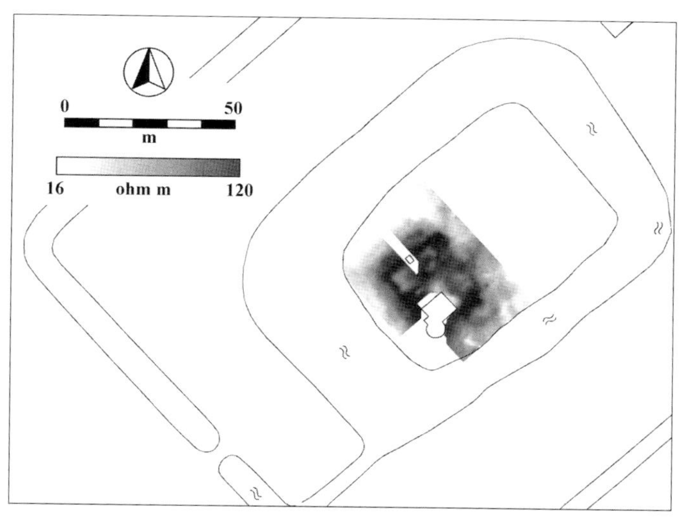

93 *Kasteel Schonauwen, Holland. The data show the brick foundations (high resistivity), with a semi-circular room in the north-west. The southern foundations connect to an existing tower. Within the north-west part of the data set can be seen a zone of low resistivity – this is evidence for a historically attested moat. In the north-east part of the survey the foundations of the bridge over the moat are visible as a zone of high resistivity. The resistance data was collected using a 1m separation Twin-Probe and converted to resistivity.* The image and information regarding this site is courtesy of RAAP

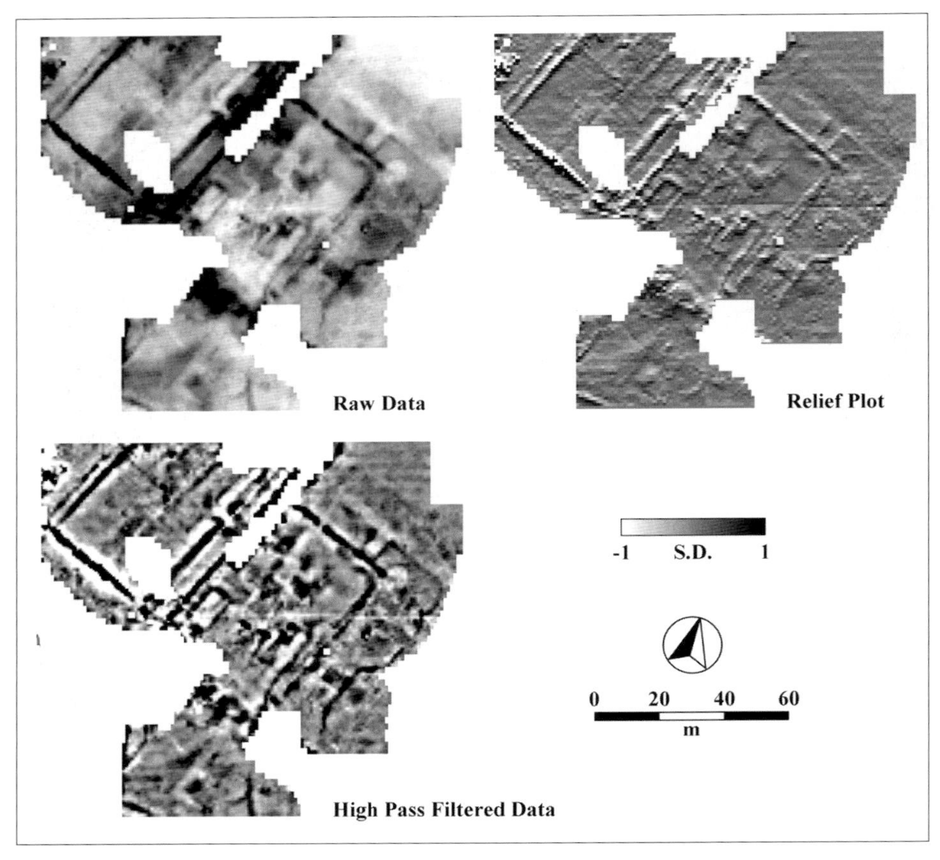

94 *Rycote Manor, Twin-probe resistance data, 1 x 1m. The complexity of the results reflects a variety of buried features including wall foundations (both surviving and robbed out), garden features and service trenches*

Gardens

The use of geophysics to investigate the layout and plans of formal gardens is a relatively recent avenue of research in archaeological geophysics. Locock (1995) has carried out a small trial on the effectiveness of dowsing in garden archaeology and seems to suggest that the method can locate metal pipes but not soil features. However, two important research papers have been written and these demonstrate that geophysical techniques can have an important role to play where the circumstances are right (Aspinall and Pocock 1995, Cole *et al.* 1997). Success seems to be dependent upon the correct selection of sampling interval, usually much finer than on normal archaeological investigations, as well as the choice of technique employed. The results from Forbury Gardens in Reading are a good example of the detail that can be resolved when looking for garden features (**95**). What is clear from the results is the remarkable detail that can be achieved of a relatively short-lived activity that produces very little in the way of significant ground disturbance.

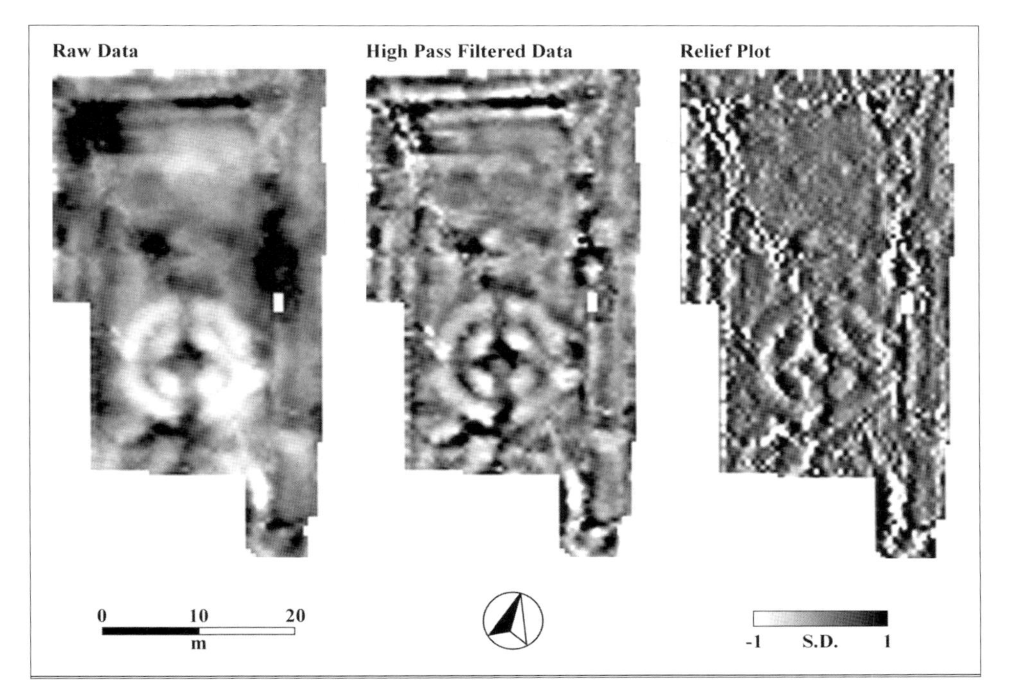

95 *Forbury Gardens, Reading. Twin-Probe resistance data collected at 0.5 x 0.5m intervals.* GSB survey for Oxford Archaeology

Agricultural features

Given that greenfield sites are often best suited to geophysical investigation, it is perhaps unsurprising that relatively recent agricultural features will leave their mark in survey data. Ridge and furrow or later ploughing, for example, can produce clear striping in datasets and while this may be of archaeological importance in its own right, often it can mask the responses from deeper archaeological features. There is often something of a dilemma when ridge and furrow is found: should it be minimised or left to dominate the response from the underlying archaeology? It is clearly part of the story of a site and it often has something to say about the potential survival of earlier deposits. However, once the extent has been noted it is often worth stripping the effect out of the data. The use of simple directional filters can overcome this problem, particularly if the alignment is known in advance so that the survey grids are aligned parallel to the ridges. If the ploughing is not visible as a topographic feature then aerial evidence may be available to help in deciding the grid orientation (**96**). We have found that in many instances where ridge and furrow itself survives as upstanding earthworks, there tends to be no magnetic contrasts because the soil that forms the ridges has simply been scooped out of the furrows and has not undergone any magnetic enhancement. By way of contrast, when the ridge and furrow is eroded by subsequent

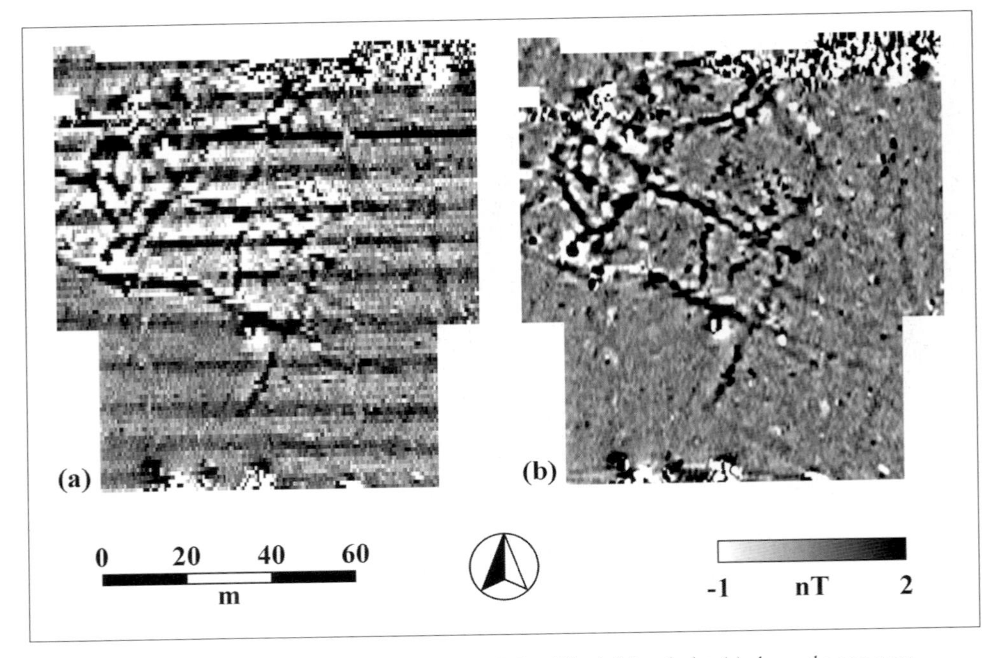

96 *Little Thetford. Fluxgate gradiometer data. 1 x 0.5m. The left-hand plot (a) shows the responses associated with former ridge and furrow cultivation, while the right-hand plot (b) show the results after a directional filter has been used to remove the anomalous responses.* GSB survey for Cambridge Archaeological Unit

ploughing, magnetic material may be incorporated in the former furrows, perhaps as a result of manuring, and thus strong contrasts become apparent. Similarly, if the ridge and furrow is overlying an earlier archaeological site, the plough furrows may cut into the magnetic deposits below and bring this material closer to the surface (**97**). Thus the ridge and furrow become more visible in the area of the pre-existing magnetically enhanced site.

Land drains have differing effects on magnetic datasets. Fired clay drains often result in linear anomalies and sometimes these are instantly recognisable by the patterns and shapes they form (**98**). Plastic drains cannot as such be detected magnetically, but they are often laid on magnetic gravels, or similar, and in these instances the narrow trenches can be detected. As with ridge and furrow, drains often bring enhanced magnetic material to the surface where they cut through pre-existing archaeological features and thus these zones of enhancement can be mapped.

Industrial features

The responses from industrial sites tend to become increasingly difficult to disentangle using geophysical techniques as we approach modern times. This is largely

97 (Above) *Fluxgate gradiometer data. 1 x 0.5m. The results show that where the ridge and furrow (R+F) cultivation overlies (and cuts into) the earlier archaeological complex, the R+F responses are themselves more strongly enhanced.* GSB survey for John Samuels Archaeological Consultants

98 (Right) *Magnetic anomalies associated with different patterns of modern land drains. Fluxgate gradiometer data at 1 x 0.5m intervals*

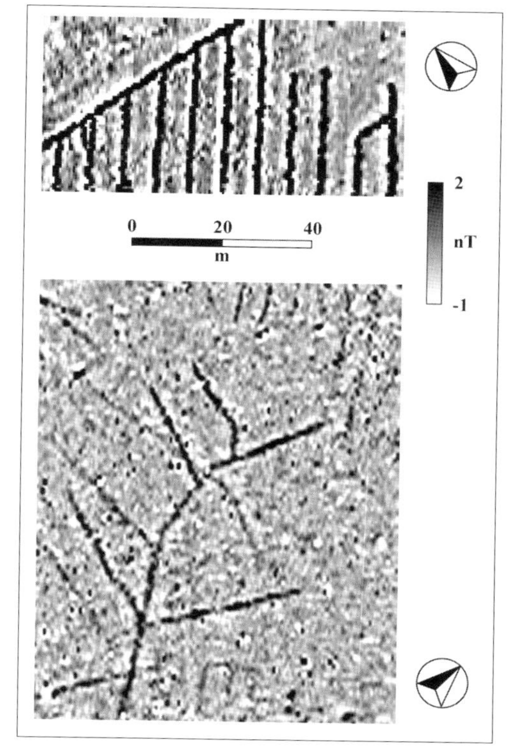

the result of the scale of the industrial enterprises and the amount of debris that they discard during the manufacturing process. However, working at a variety of scales, some authors have managed to extract an enormous amount of archaeological detail from ironworking sites. During the evaluation of whole industrial landscapes Vernon *et al.* (1998) have produced rudimentary assessment of furnace morphology. It was assumed that the response from slag material would dominate the magnetic response, which was proved not to be the case; on some of the sites features associated with iron production, e.g. furnaces, were clearly defined despite being in close proximity to slag material.

At Blaenarvon, in South Wales, we were faced with an unusual task, namely trying to find a lost viaduct. A Victorian railway, which was actually more like a tramway, was originally used to transport coal from a mine to nearby blast furnaces. It crossed a small valley but after several years this became filled with slag, from the mine, and it eventually fell into disuse. In the 1970s the valley was levelled when the site was used for landfill. Geophysical techniques were incorporated within a research design to try to relocate the line of the viaduct and to report on its survival condition. Despite attempts with electrical imaging and radar, the target proved to be too small, especially since it was buried below some 11m of highly variable overburden. We attempted to re-survey the area when some 8m had been excavated, but the results were still not convincing. However, when we surveyed over a nearby road on the edge, as opposed to the middle, of the valley the GPR transect produced very clear reflections (**99**) on the projected line of the railway. The excavation eventually uncovered an intact viaduct that had been capped to make it into a tunnel. This lined up perfectly with the radar reflections.

While the target was a known entity at Blaenarvon, we are not always faced with such a clear target. In fact, in routine evaluation work we are often only given clues as to what might lie within a proposed development area. One such case was a site where in addition to suspected Roman remains in the study zone, there was also documentary evidence for medieval mining remains in the form of possible bell pits. We adopted an approach that involved magnetic scanning followed by targeted detailed survey over suspected hotspots. Some of the results are reproduced here as they demonstrate some interesting anomalies and the difficulties of interpretation (**100**). The plots show four distinct anomalies that are displayed as a greyscale image and a 3D terrain model. The former show the strength and shape of the anomalies and these assist greatly in the interpretation of the results. The two anomalies on the left comprise quite strong (20-40nT) responses that are likely to indicate burnt or fired material, just possibly associated with small oven-type features. However, the lack of any associated archaeological-type responses in the vicinity calls into question such an interpretation; it is perhaps more likely that they are brick-capped bell-pits. By contrast, the two anomalies on the right have a much stronger negative component to the responses, indicating the presence of at least some ferrous material. While the results could always be modern in origin, the relative positions of the anomalies and the known potential of the site also

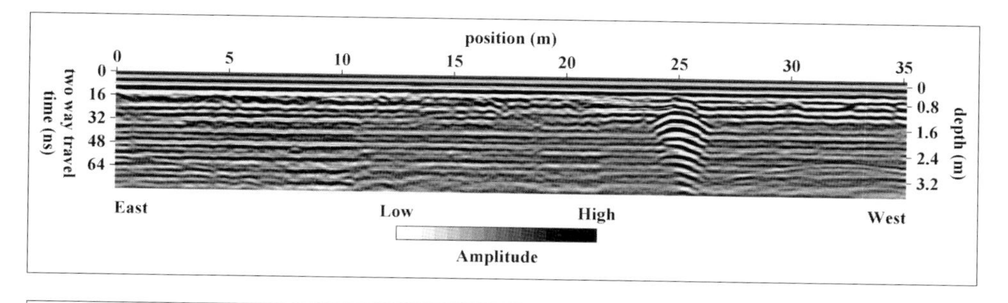

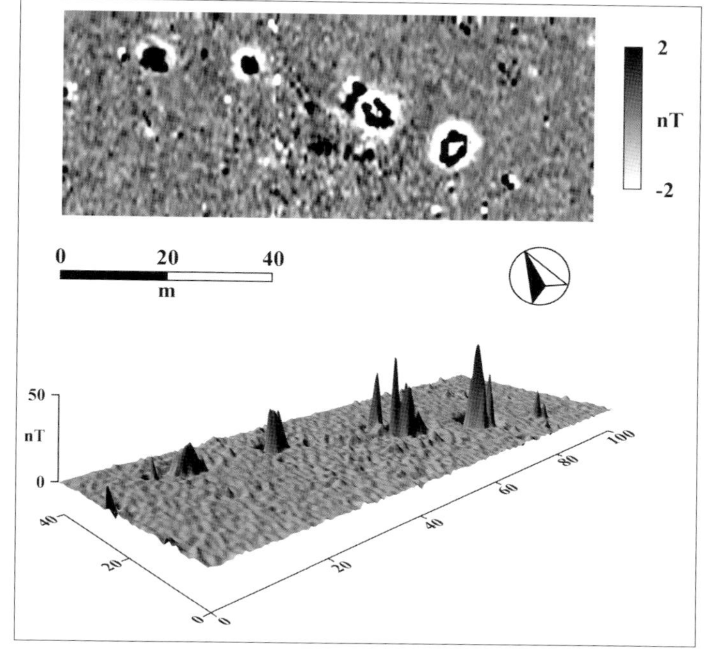

99 (Above) *GPR. A 250MHz antenna was towed along a road behind a vehicle and clearly identified a void believed to be associated with a small railway tunnel*

100 (Left) *Fluxgate gradiometer data. 1 x 0.5m. The results highlight four anomalies associated with former industrial activity, but whether they are Roman kilns or medieval or later bell-pits is impossible to answer without invasive work*

tends towards an interpretation of bell-pits, but in this instance back-filled with ferrous material. These are cases where excavation would be required to resolve the uncertainty. But having narrowed down the targets using geophysics, excavation becomes far more viable than simply relying on random trial trenching.

We have already seen that pottery kilns provide excellent targets for magnetic survey. At Tyler Hill in Kent an extensive ceramics industry is well attested; it dates from *c.*AD 1150, peaked between 1250 and 1350, and died out by about 1550, although the production of brick and peg-tile continued as late as 1900. In this case our brief was to try to locate a kiln and any ancillary structures prior to their excavation.

The field suspected of containing kilns was scanned magnetically and a strong response was noted that coincided with a marked topographic mound. The existence of a >300nT anomaly was confirmed by detailed survey and the data further indicated a double peak suggesting the presence of kiln walls surviving

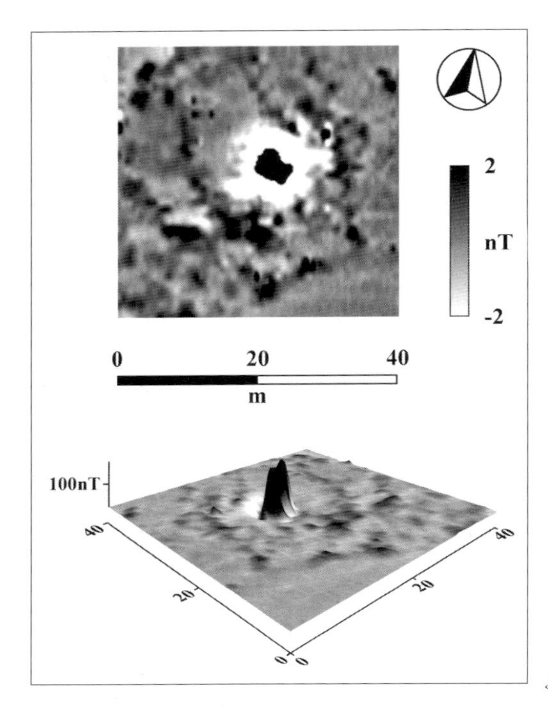

101 *Tyler Hill, Kent. Medieval tile kiln, waster dumps and workshops. Fluxgate gradiometer data. 1 x 0.5m*

intact. Surrounding the kiln was a complex of shallow ditches and beam slots that it was suggested had become backfilled with reject tiles. Some of the stronger (i.e. 5-10nT) anomalies were interpreted as the result of dumps of wasters (**101**).

Magnetic survey was followed by resistance work in an attempt to identify any detail of the kiln structure or the surrounding workshops. The kiln itself resulted in high resistance readings, but no real indications of the nature of the construction of the kiln. Other high resistance readings to the south were originally interpreted as being associated with stone foundations of ancillary buildings; it was suggested they were stone because of the lack of magnetic anomalies that would be associated with *in situ* bricks. However, on excavation the high readings were seen to reflect large pits containing concentrations of dumped waste tiles which had a greater resistance than the surrounding natural clay into which they were cut.

Although the resistance survey recorded high readings over the kiln, the lack of any structural detail led to a GPR survey being carried out as a test case. A 20m square block was centred over the kiln and transects 1m apart were investigated. The majority of the area proved to be lacking in responses as might be expected on a clay-rich subsoil. However, striking reflections were found over the kiln structure itself. The responses suggested a substantial feature surviving to a depth of approximately 1.4m. The complex nature of the reflections included 'ringing' that was indicative of a void and 'jumbled' reflections associated with walls and internal features. The excavation of the kiln explained the observed GPR responses: the intact kiln had walls standing to a height of over 1m and there was an arched roof still intact from the last firing, with a clear void below. The photograph that accompanies the GPR image reveals the inside of the kiln taken with a digital camera at the time it was first uncovered (**colour plate 21**).

Military sites

Metal detectors have often been used to locate military items that can be found in the topsoil. However, it is rare to use geophysics in a systematic manner for detecting small artefacts. An area of increasing interest for the use of geophysical techniques concerns the accurate mapping of debris associated with aircraft crash sites; the distribution of the debris can be used to understand the events that led to a particular incident. We have been involved with two instances on *Time Team* where geophysics was used alongside fieldwalking, aerial photographic study and archaeological excavation in crash investigations.

At the first site, two American B17 bombers crashed into marshes in Norfolk following a mid-air collision in 1944. One of the planes was excavated in 1976, while the other was investigated by *Time Team* in 1998. A fluxgate gradiometer and an EM61 were employed. The EM61 comprises two large rectangular coils mounted on a wheeled cart which is pulled over the ground. The operator has a large battery pack attached to their back and a hand-held data logger records the readings. An attempt was made to divide the results into ferrous and non-ferrous responses by comparing the results from the two instruments. The fluxgate only detects ferrous remains while the EM61 responds to all metal types. It was recognised that the divisions were limited because much of the debris was likely to contain both ferrous and non-ferrous components, but it meant that the non-ferrous objects could be targeted with confidence. For example, lying outside of the core area of debris an isolated EM61 response which showed as a blank in the gradiometer data proved on excavation to be part of an aileron, about 0.5m in length, made of lead and lying some 2m below the surface (**102**).

A second crash site that *Time Team* looked at was at Wierre-Effroy in France where a Spitfire came to grief in May 1940. In this instance the plane was known to have crashed into an arable field where the soils have a high clay content. This was in contrast to the Norfolk site where the ground was a soft peaty marsh, hence the spitfire was believed to be much closer to the ground surface than the B17, which negated the need for the use of the EM61. Therefore a fluxgate gradiometer was employed in conjunction with a Whites TM808 instrument (**103**), used in its metal-seeking mode. The audio output was channelled to a specially constructed data logger which converted the sound (i.e. signal strength) to a numerical value. Although this made recording of the data easier, the instrument drifted a lot and only the core of the gradiometer area could be investigated. The magnetic results identified three strong anomalies that were also pinpointed by the metal detector. Lying close together in a line, these clearly indicated the

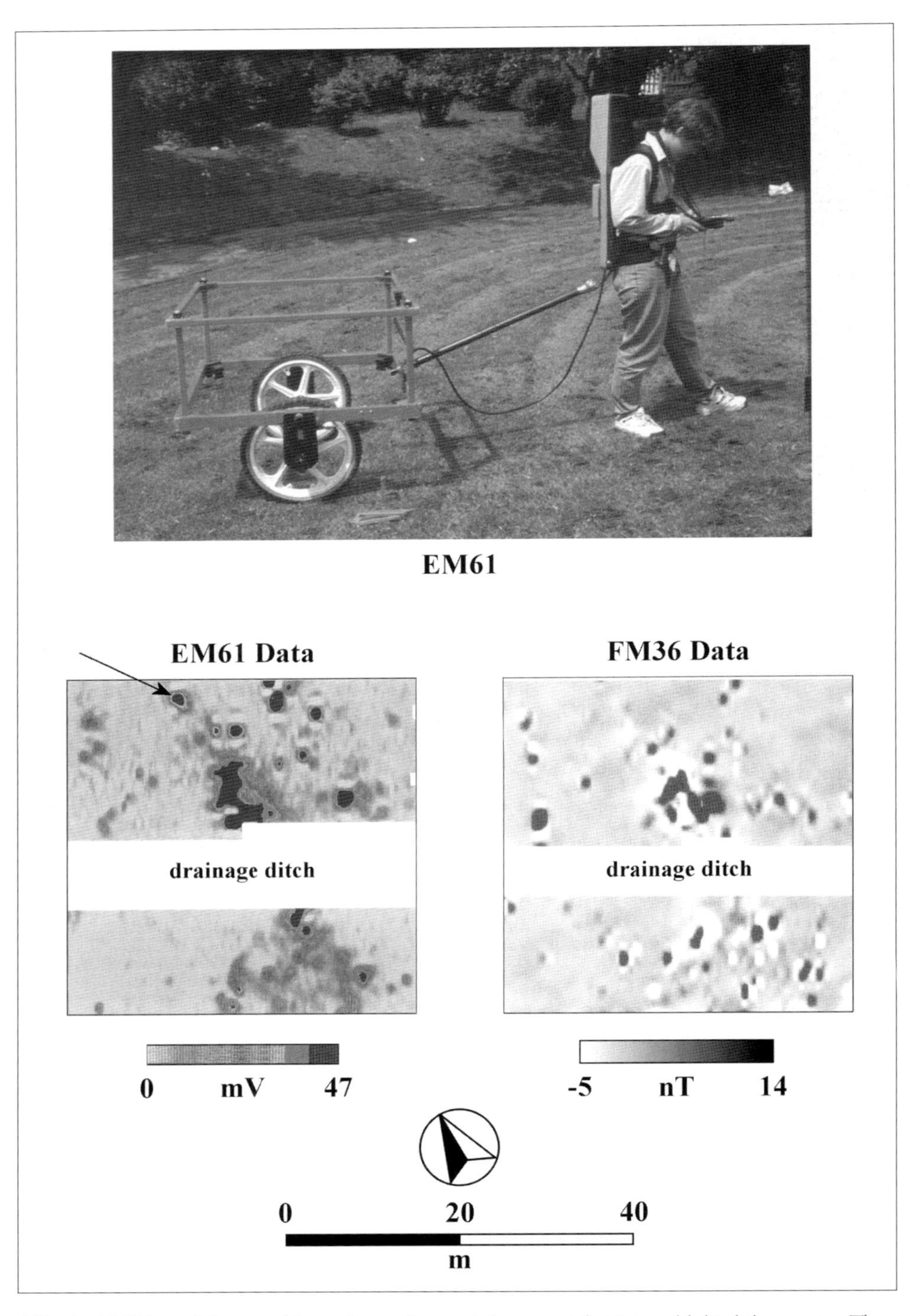

102 *An EM61 metal detector with two large coils mounted on a cart that is towed behind the operator. The batteries are stored in a backback and a data logger is carried in the hand. The EM61 data detects 'all' metal whilst the fluxgate gradiometer only sees ferrous material. The arrow marks the position of a lead aileron associated with a crashed B17 bomber aircraft found at more than 1m below the ground surface*

103 *Dr Arnold Aspinall with a Whites metal detector. The white box is a specially adapted data logger connected to the audio output of the instrument*

location of the body of the fuselage of the spitfire in the centre (**104**), flanked by two machine guns. Both instruments detected a few smaller isolated responses that related to ploughsoil debris associated with the crash. Two test pits were dug to investigate the precise origins of two anomalies and these were found to be modern agricultural machinery parts. The evidence showed that debris associated with the crash appeared to have been concentrated within *c*.10m of the main body of the aircraft, though some bits of wreckage were found in nearby field boundaries, suggesting they had been dumped there after the crash by local farmers.

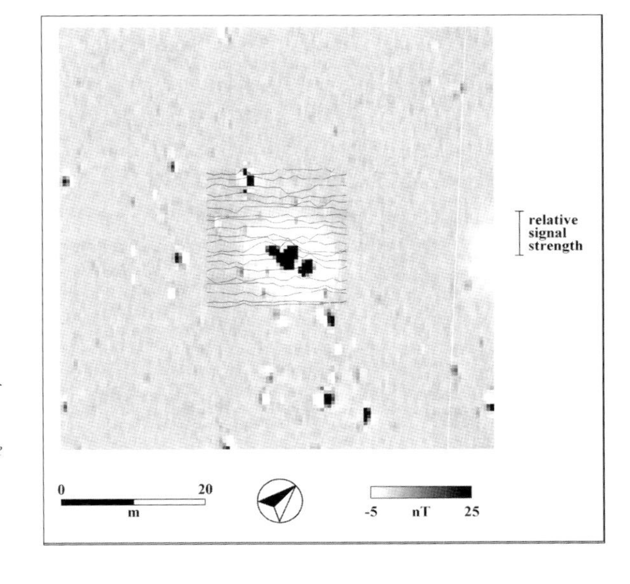

104 *A greyscale image of fluxgate gradiometer data (1 x 0.5m) indicating strong ferrous responses associated with the buried fuselage and machine guns of a Spitfire aircraft. A subsequent survey over a 20m grid, covering the core of the remains, was carried out with a Whites metal detector (1 x 1m) and is shown an X-Y plot*

Wider perspective

There are many examples from around the world that illustrate the use of geophysical techniques to prospect for later or modern archaeological sites. This in part is due to the overlap between archaeological and environmental geophysics. The latter often study the near surface, especially in urban locations. An obvious area of mutual interest involves the location of man-made cavities that not only have historic value, but may be potentially a structural engineering issue. The work of Lorenzo *et al.* (2002) produced a number of classic responses from galleries, rooms and tunnels at sites in Spain and Mexico. This was accomplished using GPR with a variety of antennae. While they conclude that a centre frequency below 100MHz was not able to locate small galleries, they still had success in locating cavities at a depth of 7m. This was achieved by analysing only individual radargrams.

An early example of time slicing of GPR data is to be found in Goodman *et al.* (1995). The data come from the Nutubaru Burial Mounds from Kyushu in Japan, a site with a date as late as AD 700, although the GPR data indicate significant anomalies from more recent periods. The authors use the slices to find a previously unknown burial with a ditch approximately 22m across. Within the ring is an anomaly that is believed to be a collapsed burial chamber. In the published article the authors make mention of very near surface features that can be imaged in the time slices but cannot be seen in the individual traverses. Since that article, this observation has been commonly reported and time slicing is now regarded as the best way to extract coherent data near the surface of the ground. The 300MHz GPR data have been re-processed by Goodman and can be seen in **colour plate 22** and the reader is referred to the original article to see how much display technology has advanced since 1995.

The use of GPR inside modern towns or buildings to look for archaeology has become fairly routine, although the results are often variable. While a large number of the early surveys were speculative, many of the more recent publications are focused on particular archaeological problems. For example there are several articles that document survey around or within medieval churches. The fact that there are often obstacles in the survey area means that the strategies are usually a mixture of individual transects and very close traverse work for time slice production or 3D visualisation (see Leucci 2002). The work of Leckebusch (2000) in this field is particularly important. In that article he discusses the problems associated with the visualisation and interpretation of 3D GPR information and concludes that neither individual traverses nor time slices can allow a complete understanding of the subsurface. As an alternative he describes a volume representation that is the result of 'semi-automatic picking of reflections and displayed in a 3D perspective view with shaded relief' (Leckebusch 2000, p.198). This potentially allows the representation and, crucially, interpretation of finer archaeological detail (**colour plate 24**).

Summary points

• Due to the diversity of the archaeology in recent times, no one geophysical technique has preference, though with speed and money always being a major consideration, especially in evaluation work, resistance survey, magnetometry and GPR overshadow the use of other techniques. Sometimes the archaeology is so distinctive, for example on industrial sites, that only one technique may be required.

• Given the extensive range of sites, it is imperative that all documentary sources are consulted, for example old maps and plans and archaeological archives. While this consultation does not have to be specifically carried out by the geophysicist, they should be aware of the information when interpreting results.

• Many later historic sites will by their nature occupy urban and semi-urban locations and as such there is a greater likelihood of modern pipes and conduits being present. Local authorities and utility providers carry maps and plans of these services, though the accuracy cannot always be relied upon.

• While surveys on rural sites can often follow similar approaches, more specialised strategies are often required on later historic sites, where the archaeology tends to be more complex and deep stratigraphy is often encountered. Much more time and effort is required in choosing the appropriate technique and devising appropriate sampling intervals.

9

LOOKING FORWARD TO PROSPECTING THE PAST

In this concluding chapter we consider some of the future challenges faced by archaeological geophysicists. Suffice to say that in the past geophysicists have tended to adapt to questions posed by the archaeological community and they will continue to address these needs in the future. However, despite the fact that innovation will continue at a fast pace, the Universal Ditch Detector is not going to materialise. Improvements in processing speeds and the quality of analytical software may mean 'machine' interpretation of geophysical data will soon reach an acceptable level, especially when multi-sensors or multi-techniques are used.

Within the last decade the instrumentation and methodologies associated with the three main geophysical techniques (magnetic, resistance and GPR) have improved tremendously. Magnetic surveys in particular have become the first option due to the ability to cover the ground quickly with this technique. The range in commercial magnetic prospecting systems that is now available is the best ever; on the one hand are fluxgate systems with differing spaced sensors and dual data capture instruments and on the other are wheeled CV systems with banks of sensors.

The dominance of the Twin-Probe array in resistance surveys is surprising, given the wide range of other configurations available, and may even be to the detriment of prospecting in archaeology. Even though considerable primary research was undertaken on the various probe arrays in the 1960s and 1970s, the reason why many were not implemented was due to practical and technical rather than theoretical reasons. With the increased sophistication of field equipment it is time to re-analyse the 'four probes good, two probes better' mantra that has dominated resistance survey. Even Arnold Aspinall, the doyen of the Twin-Probe, has appealed for a step back (Aspinall and Gaffney 2001). It is time to review some of the other probe arrays, such as the Square and the Schlumberger. After sixty years of resistance survey, there is still a need for more information on 'probe geometry' on the archaeological scale.

In terms of implementation the Twin-Probe configuration reduces survey times considerably compared to early Wenner surveys. However, there has been little improvement since the introduction of the multiplex system developed by Geoscan Research. Trials with automated wheeled systems have been carried out intermittently over the years and this is an area of research where advances are likely to be made. French researchers have recently reported progress in an area in which they have been working for more than 30 years; they have produced a traditional probe system that can be towed behind a vehicle as well as an electrostatic version that can be used on hard surfaces (Pannisod *et al.* 1998). At the time of writing Geoscan Research are finalising a new wheeled resistance cart.

In GPR the most likely advancement in hardware will be the use of multi-frequency antennae. While these instruments already exist, the cost tends to be prohibitive when it comes to archaeological investigations. On the software side, emphasis will continue to be on the production of time slices and 3D volume visualisations, though more emphasis should be given to the study and interpretation of radargrams and the modelling of reflections. Due to the complexity of response within GPR data it is imperative that tests are made on simple archaeological or pseudo-archaeological sites built for technological evaluation, e.g. Hildebrand *et al.* (2002), or within scaled down laboratory experiments, e.g. Leckebusch and Peikert (2001). Given the closeness of seismics to GPR there may well be advances in this field of study which have tended to be neglected up until now.

It is perhaps ironic that one of the biggest limiting factors in the speed at which geophysical surveys are carried out is connected to the relatively slow speed of conventional surveying techniques. The time spent establishing and tying in a survey grid can be similar to the time spent collecting data on a small survey. Yet due to the unpredictable nature of archaeological remains, a rigid survey grid and inter-grid markers will always be required when carrying out area geophysical investigations. Random data collection will not work when investigating the detail of archaeological sites. It is true that certain types of archaeological sites may follow particular plans but there is no guarantee that they will duplicate other known sites; this fact might only be established as a result of systematic data collection strategies.

GPS has helped speed up the setting out process, but at present it is often still necessary to have independent tie-ins so that others can relocate the survey grid without resorting to satellite technology. Future developments of cheap, accurate, hand-held GPS systems will render taped tie-ins unnecessary. Although some geophysical manufacturers have attempted to append GPS onto their equipment, the relative high cost of the GPS has meant that few workable systems exist. As GPS technology decreases in price by comparison to geophysical equipment, it is more likely that fully integrated systems will appear. This will be particularly welcome with regard to scanning devices such as fluxgate gradiometers and magnetic susceptibility coils. The ability to locate anomalies of potential interest directly onto digital maps, thus overcoming the necessity to tie-in scanned anomalies, will be very welcome for all.

Unfortunately, the ability to set out survey grid substantially faster than it takes at present is unlikely to happen in the foreseeable future. Faced with this knowledge, archaeological geophysicists have looked at ways of speeding up data collection. We have already discussed the results of hand-held dual gradiometer systems in Britain and referred to the wheeled multi-probe systems used in Austria and Germany. While banks of multiple magnetic sensors present many technical obstacles, e.g. matching sensors in terms of noise and developing software to balance out the differences, these problems have been largely sorted. However, the question of variable topography is more difficult to solve.

There is a divide between those who wish to collect ever-increasing sample densities and those who simply wish to cover larger survey areas. Whilst these objectives are not necessarily incompatible, the archaeological geophysicist must appreciate that there is a difference between the best possible geophysical data set that will produce the 'truest' image and the needs of the archaeologist for accurate geophysical information. That line needs to be drawn in the sand; to cross it may be seen as an attempt to obviate the need for the spade and the trowel. To simplify the problem, is it necessary for a geophysical survey to locate the smallest of features, such as post-holes, or, is it sufficient to define the nature and extent of an archaeological site? Although we can aspire to the first, we must be prepared to fail – the detail that the human eye can discern with the aid of the trowel will always be considerably better than any remotely sensed data. The future role for geophysics must be to problem-solve within focused research strategies.

Given the need for faster survey speeds and the ever increasing size of the datasets, it is somewhat surprising that issues such as data transfer times from instruments to computers have only recently been addressed by instrument designers. In view of the preponderance of USB and Bluetooth connections and the whole gamut of compact digital storage devices that exist in the wider world, it is perplexing that most geophysical instruments still rely on RS232 connectors for data transfer. Hopefully manufacturers of future geophysical instruments will address this issue.

Despite the massive size of the data sets, the storage capabilities and processing powers of field computers are fortunately keeping pace with the demands. The ability to handle large datasets remains vital and dwarfs what was achieved in 1963 when the first computerised analysis of magnetic data was carried out (see Scollar 1990). This aspect of survey, i.e. data processing, will continue to be a growth area in research. However, it should be remembered that archaeological sites, although often highly structured in terms of use, often exhibit apparently random distributions of small features and simply using standard algorithms to eliminate random noise may induce rather than alleviate problems (Aspinall 1992).

The interpretation of geophysical data in archaeological terms is severely hindered by the lack of 'feedback' from sites that have been surveyed and subsequently excavated (Boucher 1996). It is imperative that excavators tell the geophysicist of the outcome of the invasive work. In practice, the archaeologist is often quick to point out the deficiencies in an interpretation, but it is vital that positive feedback is offered, if the level of confidence in interpretation is to improve. Whenever possible, information should be shared among the archaeological community and we have already mentioned in chapter 5 the English Heritage database and the information held by the Archaeology Data Service. The latter has issued guidelines on the digital preservation of geophysical data (Schmidt 2002) though some of the practical aspects are still unclear.

As we discussed in the preface, archaeological geophysics as a niche subject has probably received more media attention than any comparable scientific discipline, a fact attributed entirely to the *Time Team* television series. The programme has inspired a generation of young archaeological geophysicists who are bringing new skills, ideas and talents to the subject. Archaeology is gradually throwing off its musty image and engendering interest in an increasingly wide range of people. Whilst the geophysics used in archaeology and discussed in this book may not be 'rocket science' for many, it is science that is challenging, novel, exciting and achievable.

We have been fortunate to be involved in a period of rapid development and implementation of archaeological geophysics in Britain. We hope that others will continue to be inspired by this work to take on future research and investigations. But the geophysicists of the future should never lose sight of the bigger picture. It must be recognised, for example, that the circular magnetic anomaly they have surveyed is more than an aberration in the Earth's magnetic field that can be measured with great precision; it would be sad if revealing the buried past became divorced from understanding the past.

BIBLIOGRAPHY

Aitken, M.J. 1958 'Magnetic Prospecting. 1 – The Water Newton Survey' *Archaeometry* 1(1) 24-9.

Aitken, M.J. 1974 *Physics and Archaeology* (2nd edn) Clarendon Press. Oxford.

Aitken, M.J. 1986 'Proton Magnetometer Prospection: Reminiscences of the First Year' *Prospezioni Archaeologiche* 10 15-17.

Aitken, M.J. and Tite, M.S. 1962 'Proton Magnetometer Surveying on Some British Hillforts' *Archaeometry* 5 126-34.

Alldred, J.C. 1964 'A fluxgate gradiometer for archaeological surveying' *Archaeometry* 7 14-19.

Annan, A.P. 1997 *Ground-Penetrating Radar: Workshop Notes* Sensors & Software. Mississauga. Canada.

Aspinall, A. 1992 'New Developments in Geophysical Prospection' in A.M. Pollard (ed). *New Developments in Archaeological Science* Proceedings of the British Academy 77 (1992) Oxford University Press for the British Academy. 233-44.

Aspinall, A. and Crummett J.G. 1997 'The Electrical Pseudosection' *Archaeological Prospection* 4 37-48.

Aspinall, A. and Gaffney, C.F. 2001 'The Schlumberger Array – Potential and Pitfalls' *Archaeological Prospection* 8(3) 199-209.

Aspinall, A. and Haigh, J.G.B. 1988 'A review of techniques for the graphical display of geophysical data' in S.P.Q. Rahtz (ed.) *Computer and Quantitative Methods in Archaeology* British Archaeological Reports, International Series 446 (1988) Oxford. 295-307.

Aspinall, A. and Lynam, J.T. 1970 'An induced polarisation instrument for the detection of near surface features' *Prospezioni Archeologiche* 5 67-75.

Aspinall, A. and Pocock, J. 1995 'Geophysical Prospection in Garden Archaeology: an appraisal and critique based on case studies' *Archaeological Prospection* 2(2) 61-84.

Aston, M.A., Martin, M.H. and Jackson, A.W. 1998 'The potential for heavy metal soil analysis on low status archaeological sites at Shapwick, Somerset' *Antiquity* 72 838-47.

Atkin, M. and Milligan, R. 1992 'Ground-Probing Radar in Archaeology – Practicalities and Problems' *The Field Archaeologist* 16 288-91.

Atkinson, R.J.C. 1952 'Methodes electriques de prospection en archeologie' in A. Laming (ed.) *La Decouverte du Passe* (1952) A. and J. Picard. Paris. 59-70.

Atkinson, R.J.C. 1953 *Field Archaeology* (2nd edition) Methuen. London

Bailey, R.N., Cambridge, E. and Briggs, H.D. 1988 *Dowsing and Church Archaeology* Intercept. Winbourne.

Barker, R.D. 1992 'A simple algorithm for electrical imaging the subsurface' *First Break* 10 53-62.

Becker, H. 1995 'From Nanotesla to Picotesla – A New Window for Magnetic Prospecting in Archaeology' *Archaeological Prospection* 2(4) 217-28.

Becker, H. 1996 *Archaologische Prospektion, Luftbildarchaologie und Geophysik* Bayerisches Landesamt fur Denkmalplege. Munchen.

Bellerby, T.J., Noel, M., and Brannigan, K. 1990 'A thermal method for archaeological prospection: Preliminary investigations' *Archaeometry* 32 191-203.

Bettess, F. 1992 *Surveying for Archaeologists* (revised edition). University of Durham. Durham.

Bevan, B.W. 1995 'Geophysical Prospecting' *American Journal of Archaeology* 99 88-90.

Bevan, B.W. 2000 'An early Geophysical Survey at Williamsburg, USA' *Archaeological Prospection* 7 51-8.

Bevan, B.W. and Kenyon, J. 1975 'Ground-Penetrating Radar for Historical Archaeology' *Masca Newsletter* 11(2) 2-7.

Boucher, A.R. 1996 'Archaeological Feedback in Geophysics' *Archaeological Prospection* 3(3) 129-40.

Bowden, M. 1999 *Unravelling the Landscape. An inquisitive approach to archaeology* Tempus. Gloucestershire.

Buteux, S., Gaffney, V., White, R. and van Leusen, M. 2000 'Wroxeter Hinterland Project and Geophysical Survey at Wroxeter' *Archaeological Prospection* 7(2) 69-80.

Catherall, P.D., Barnett, M. and McClean, H. 1984 *The Southern Feeder* British Gas Corporation. London.

Cave-Penny, H. 1995 'Time Team and the Saxon Cemetery at Winterbourne Gunner' *The Field Archaeologist* 23 6-7.

Chamberlain, A.T., Sellars, W., Proctor, C. and Coard, R. 2000 'Cave Detection in Limestone using Ground Penetrating Radar' *Journal of Archaeological Science* 27 957-64.

Clark, A.J. 1973 *Stones of Stenness, Orkney: Magnetometer Survey* Ancient Monuments Laboratory (Old Series) No. 1613. Unpublished.

Clark, A.J. 1996 *Seeing Beneath the Soil* (2nd edition). Batsford. London.

Clark, A.J. and Haddon-Reece, D. 1973 'An automated recording system using a Plessey fluxgate gradiometer' *Prospezioni Archaeologiche* 7-8 101-13.

Colani, C. and Aitken, M.J. 1966 'A New Type of Locating Device. II – Field Trials' *Archaeometry* 9 9-19.

Cole, M.A., David, A.E.U., Linford, N.T., Linford, P.K. and Payne, A.W. 1997 'Non-destructive techniques in English gardens: geophysical prospecting' *Journal of Garden History* 17 26-39.

Conyers, L.B. and Goodman, D. 1997 *Ground-Penetrating Radar: an introduction for archaeologists.* AltaMira Press. Walnut Creek.

Conyers, L.B., Ernenwein, E.G. and Bedal, L. 2002 Ground Penetrating Radar (GPR) 'Mapping as a Method for Planning Excavation Strategies, Petra, Jordan' *E-tiquity* 1 e-tiquity.saa.org/~etiquity/title1.html.

Coppack, G. 1990 *Abbeys and Priories* Batsford/English Heritage. London.

Coppack, G. and Aston, M. 2002 *Christ's Poor Men. The Carthusians in England* Tempus. Gloucestershire.

Corney, M., Gaffney, C.F. and Gater, J.A. 1994 'Geophysical Investigations at the Charlton Villa, Wiltshire (England)' *Archaeological Prospection* 1(2) 121-8.

Dabas, M., Hesse, A. and Tabbagh, J. 2000 'Experimental Resistivity Survey at Wroxeter Archaeological Site with a Fast and Light Recording Device' *Archaeological Prospection* 7(2) 107-18.

Darlington, J. and Shiel, D. 2001 'Geophysical Survey at Stafford Castle' in J. Darlington (ed.) *Stafford Castle: Survey, Excavation and research 1978-1998. Volume I – the Surveys* (2001) Stafford Borough Council. Stafford. 117-46.

David, A. 1995 *Geophysical Survey in Archaeological Field Evaluation* English Heritage Research and Professional Services Guideline No. 1. London.

David, A. and Payne, A. 1997 'Geophysical Surveys within the Stonehenge Landscape: A Review of Past Endeavour and Future Potential' *Proceedings of the British Academy* 92 73-113.

Dawson, M. and Gaffney, C. 1995 'The Application of Geophysical Techniques within a Planning Application at Norse Road, Bedfordshire (England)' *Archaeological Prospection* 2 103-15.

Dearing, J. 1999 'Magnetic Susceptibility' in F. Foldfield, J. Smith, J. Walden (eds) *Environmental Magnetism: A Practical Guide* (1999) Quaternary Research Association, Technical Guide Number 6 35-62.

Dockrill, S. forthcoming *Toftsness – an archaeological landscape* Historic Scotland.

Dobbs, C.A., Maki, D.L. and Forsberg, M. 1997 'The Use of Ground-Penetrating Radar on Small Prehistoric Sites in the Upper Midwestern United States' in L. Dingwall, S. Exon, V. Gaffney, S. Laflin and M. van Leusen (eds) *Archaeology in the Age of the Internet: Proceedings of the 25th Anniversary Conference of CAA* (1997) Birmingham.

DoE 1990 *Planning Policy Guidance 16: Archaeology and Planning* HMSO.

Esmonde Cleary, S. 1999 'Roman Britain: Civil and Rural Society' in J.R. Hunter and I. Ralston (eds) *The Archaeology of Britain* (1999) Routledge. London. 157-75.

Fassbinder, J.W.E., Stanjek, H. and Vali, H. 1990 'Occurrence of magnetic bacteria in soil' *Nature* 343 161-3.

Fisher, P.M. 1980 'Applications of Technical Devices in Archaeology – the Use of X-rays, Microscope, Electrical and Electromagnetic Devices and Subsurface Interface Radar' *Studies in Mediterranean* LXIII.

Fowler, P. 1959 'Magnetic Prospecting: An Archaeological Note about Madmarston' *Archaeometry* 2 35-9.

French, C. forthcoming *Prehistoric landscape development and human impact in the upper Allen valley, Cranborne Chase, Dorset.*

Frohlich, B. and Lancaster, W.J. 1986. 'Electromagnetic surveying in current Middle Eastern archaeology: Application and evaluation' *Geophysics* 51(7) 1414-25.

Fulford, M.G. and Allen, J.R.L. 1992 'Iron -Making at the Chesters Villa, Woolaston' *Britannia* XXIII.

Gaffney, C.F. and Gater, J.A. 1992 'The Geophysical Survey' in M.G. Fulford and J.R.L. Allen, 'Iron-Making at the Chesters Villa, Woolaston' *Britannia* XXIII 162-5.

Gaffney, C.F. and Gater, J.A. 1993 'Development of Remote Sensing. Part 2. Practice and method in the application of geophysical techniques in archaeology' in J.R. Hunter and I. Ralston (eds) *Archaeological Resource Management in the UK* Alan Sutton. Stroud.

Gaffney, C.F., Gater, J.A., Linford, P., Gaffney, V. and White, R. 2000 'Large-scale Systematic Fluxgate Gradiometry at the Roman City of Wroxeter' *Archaeological Prospecting* 7(2) 81-100.

Gaffney, C.F., Gater, J.A. and Ovenden, S.M. 2002 *The use of geophysical Techniques in Archaeological Evaluations*. Institute of Field Archaeologists Paper No. 6.

Gater, J.A. 1981 *Hadrian's Wall Resistivity Survey* Geophysics Report 24/1981. Ancient Monuments Laboratory. Unpublished.

Gater, J., Leech, R., and Riley, H. 1993 'Later Prehistoric and Romano-British Settlements in South Somerset: Some Recent Work'. *Proceedings of the Somerset Archaeological and Natural History Society* 137 41-58.

Goodman, D., Nishimura, Y. and Rogers, J.D. 1995 'GPR Time Slices in Archaeological Prospection, *Archeological Prospection* 2(2) 85-9.

Goulty, N.R. and Hudson, A.L. 1994 'Completion of the seismic refraction survey to locate the vallum at Vindobala, Hadrian's Wall' *Archaeometry* 36 327-35.

Griffiths, D.H. and Barker, R.D. 1994 'Electrical Imaging in Archaeology' *Journal of Archaeological Science* 21 153-8.

GSB 2002 *Geophysical Survey Report: Orkney World Heritage Site*. Report No. 02/61 GSB Prospection. Unpublished.

Hall, M.B. 1974 *Foundations Unearthed* (4th edition) Veritas Press. Los Angeles CA.

Hanson, W.S. 1999 'Roman Britain: Military Dimension' in J.R. Hunter and I. Ralston (eds) *The Archaeology of Britain* Routledge. London. 135-56.

Harding, P. and Lewis, C. 1997 'Archaeological Excavations at Tockenham, 1994' *Wilts. Arch. N. Hist. Mag* 90 29-41.

Haselgrove, C. 1999 'The Iron Age' in J.R. Hunter and I. Ralston (eds) *The Archaeology of Britain* Routledge. London. 113-34.

Herbich, T. 1993 'The method of estimation of the extent of the mining field of flint mines through observation of the arrangement of surface layers' *Archeologia Polski* XXXVIII 23-35.

Heron, C.P. and Gaffney, C.F. 1987 'Archaeogeophysics and the site: ohm sweet ohm?' in C.F. Gaffney and V.L. Gaffney (eds) *Pragmatic Archaeology: Theory in Crisis?* British Archaeological Report 167 71- 81.

Hesse, A. 1966 *Prospections geophysiques a fiable profondeur. Applications a l'Archeologie* Dunod. Paris.

Hesse, A. 2000 'Count Robert du Mesnil du Buisson (1895-1986), A French Precursor in Geophysical Survey for Archaeology' *Archaeological Prospection* 7 43-9

Hildebrand, J.A., Wiggins, S.M., Henkart, P.C. and Conyers, L.B. 2002 'Comparison of Seismic Reflection and Ground-Penetrating Radar Imaging at the Controlled Archaeological Test Site, Champaign, Illinois' *Archaeological Prospection* 9(1) 9-22.

Howell, M 1966 'A Soil Conductivity Meter' *Archaeometry* 9 20-23.

Kearey, P. and Brooks, M. 1991 *An introduction to geophysical exploration* (2nd edition) Blackwell Scientific. Oxford.

Kelly, M.A., Dale, P. and Haigh, J.G.B. 1984 'A microcomputer system for data logging in geophysical surveying' *Archaeometry* 2 183-91.

Kvamme, K. 2003 'Multidimensional Prospecting in North American Great Plains Village Sites' *Archaeological Prospecting* 10 (2) 131-42.

Laming, A. (ed.) 1952 *La decouverte du passe* A. and J. Picard. Paris.

Le Borgne, E. 1955 'Susceptibilité magnétiques anomale du sol superficial' *Annales de Geophysique* 11 399-419.

Le Borgne, E. 1960 'Influence du feu sur les proprieties magnetiques du sol et du granite' *Annales de Geophysique* 16 159-95.

Leckebusch, J. 2000 'Two and Three-dimensional Ground-penetrating radar Surveys Across a Medieval Choir: A case study in Archaeology' *Archaeological Prospecting* 7(3) 189-200.

Leckebusch, J. and Peikert, R. 2001 'Investigating the True Resolution and Three-Dimensional Capabilities of Ground-Penetrating Radar Data in Archaeological Surveys: Measurements in Sand Box' *Archaeological Prospection* 8(1) 29-41 .

Leucci, G. 2002 'Ground-penetrating Radar Survey to Map the Location of Buried Structures under Two Churches' *Archaeological Prospection* 9(4) 217-28.

van Leusen, M. 1998 'Dowsing and Archaeology' *Archaeological Prospection* 5(3) 123-38.

Linford, N.T. 1998 'Geophysical survey at Boden Vean, Cornwall, including an assessment of the microgravity technique for the location of suspected archaeological void features' *Archaeometry* 40(1) 187-216.

Linford, N.T. and Canti, M.G. 2001 'Geophysical Evidence for fires in Antiquity: Preliminary Results from an Experimental Study' *Archaeological Prospection* 8(4) 211-26.

Locock, M. 1995 'The Effectiveness of Dowsing as a method of Determining the Nature and location of Buried Features on Historic Garden Sites' *Archaeological Prospection* 2(1) 15-18.

Lorenzo, H., Hernandez, M.C. and Cuellar, V. 2002 'Selected Radar Images of Man-Made Underground Galleries' *Archaeological Prospection* 9(1) 1-8.

Lyall J. and Powlesland, D. 1996 *The application of high resolution fluxgate gradiometery as an aid to excavation planning and strategy formulation* http://intarch.ac.uk/journal/issue1/lyall_index.html

Maher, B.A. and Taylor, R.M. 1988 'Formation of ultrafine grained magnetite in soils' *Nature* 336 368-70.

Marriot, J. and Yarwood, R. 1988 'A Prehistoric Earthwork at Altofts, West Yorkshire' Yorkshire SMR. Unpublished.

Martin, L 2000. 'Silchester Roman Town, Hampshire. Report on Geophysical Survey, March 2000'. Geophysical Report No 65/2000 Ancient Monuments Laboratory. Unpublished.

Musset, A.E. and Khan, M. 2000 *Looking Into The Earth: An Introduction to Geological Geophysics* Cambridge University Press. Cambridge.

Neighbour, T., Strachan, R. and Hobbs, B.A. 2001 'Resistivity Imaging of the Linear Earthworks at the Mull of Galloway, Dumfries and Galloway' *Archaeological Prospection* 8(3) 157-62.

Neubauer, W. 2001 *Magnetische Propektion in der Archaologie* Wien.

Neubauer, W., Eder-Hinterleitner, A., Seren, S. and Melichar, P. 2002 'Georadar in the Roman Civil Town Carnuntum, Austria: An Approach for Archaeological Interpretation of GPR Data' *Archaeological Prospection* 9 135-56.

Nishimura, Y. and Goodman, D. 2000 'Ground-penetrating Radar Survey at Wroxeter' *Archaeological Prospection* (2) 101-6.

Noel, M. and Biwen, X. 1992 'Cave Detection using Electrical Resistivity Tomography' *Cave Science* 19(3) 91-4.

Noel, M. and Xu, B. 1991 'Archaeological investigation by electrical resistivity tomography: a preliminary study' *Geophysical Journal International* 107 95-102.

Ovenden, S.M. 1994 'Application of Seismic Refraction to Archaeological Prospecting' *Archaeological Prospecting* 1(1) 53-64.

Panissod, G., Debas, M., Florsch, N., Hesse, A., Jolivet, A., Tabbagh, A. and Tabbagh, J. 1998 'Archaeological Prospecting using Electric and Electrostatic Mobile Arrays' *Archaeological Prospection* 5(4) 239-52.

Payne, A. 1996 'The Use of Magnetic Prospection in the Exploration of Iron Age Hillfort Interiors in Southern England' *Archaeological Prospection* 3 163-84.

Philpot, F.V. 1973 'An improved fluxgate gradiometer for archaeological surveys' *Prospezioni Archaeologiche* 7-8 99-105.

Piro, S., Goodman, D. and Nishimura, Y. 2003 'The Study and Characterisation of Emperor Traiano's Villa (Altopiani di Arcinazzo, Roma) using High-resolution Integrated Geophysical Surveys' *Archaeological Prospection* 10(1) 1-26.

Pringle, J.K., Westerman, A.R., Schmidt, A., Harrison, J., Handley, D., Beck, J., Donahue, R.E., and Gardiner, A.R. 2002 'Investigating Peak Cavern, Castleton, Derbyshire,UK: Integrating cave survey, geophysics, geology and archaeology to create a 3D digital CAD model' *Cave and Karst Science* 29(2) 67-74.

Sarris, A., Athanassopoulou, E., Doulgeri-Intzessiloglou, A., Skafida, Eu and Weymouth, J. 2002 'Geophysical Prospection Survey of an Ancient Amphorae Workshop at Tsoukalia, Alonnisos (Greece)' *Archaeological Prospection* 9(4) 183-96.

Schmidt, A. 2002 *Geophysical Data in Archaeology: A Guide to Good Practice* Archaeology Data Service. Oxbow. Oxford. (http://ads.ahds.ac.uk/project/goodguides/geophys/)

Scollar, I. 1965 'Recent developments in magnetic prospecting in the Rheinland' *Prospezione Archaeologiche* 1 43-50.

Scollar, I. 1990 *Archaeological Prospecting and Remote Sensing* Cambridge University Press. Cambridge.

Schleifer, N., Weller, A., Schneider, S. and Junge, A. 2002 'Investigation of a Bronze Age Plankway by Spectral Induced Polarisation' *Archaeological Prospection* 9 243-53.

Schwarz, G.T. 1961 'The "Zirkelsonde": A New Technique for Resistivity Surveying' *Archaeometry* 4 67-70.

Slepak, Z. 1999 'Electromagnetic Sounding and High-precision Gravimeter Survey Define Ancient Stone Building Remains in the Territory of Kazan Kremlin (Kazan, Republic of Tatarstan, Russia)' *Archaeological Prospection* 6(3) 147-60.

Sowerbutts, W.T.C. 1988 'The use of geophysical methods to locate joints in underground metal pipelines' *Quarterly Journal of Engineering Geology* 21 273-81.

Sowerbutts, W.T.C. and Mason, R.W. 1984 'A microcomputer-based system for small-scale geophysical surveys' *Geophysics* 49 189-93.

Spoerry, P (ed.) 1992 *Geoprospection and the Archaeological Landscape* Oxbow Monographs 18. Oxford.

Stove, G.C. and Addyman, P.V. 1989 'Ground Probing Impulse Radar: An Experiment in Archaeological Remote Sensing at York' *Antiquity* 63 337-42.

Szymanski, J.E. and Tsourlos, P. 1993 'The resistivity tomography technique for archaeology: an introduction and review' *Archaeologia Polona* 31 5-32.

Tabbagh, A. 1986 'Applications and advantages of the Slingram electromagnetic method for archaeological prospecting' *Geophysics* 51 576-84.

Telford, W.M., Geldart, G.P. and Sheriff, R.E. 1990 *Applied Geophysics* (2nd edition) Cambridge University Press. Cambridge.

Thompson, R. and Oldfield, F. 1986 *Environmental Magnetism* Allen and Unwin. London.

Tite, M.S. 1972 *Methods of Physical Examination in Archaeology* Seminar Press. London and New York.

Tite, M.S. and Mullins, C. 1970 'Electromagnetic prospecting on archaeological sites using a soil conductivity meter' *Archaeometry* 12 97-104.

Tite, M.S. and Mullins, C. 1971 'Enhancement of the magnetic susceptibility of soils on archaeological sites' *Archaeometry* 13 209-19.

Vickers, R.S., Dolphin, L.T. and Johnson, D. 1976 'Archaeological Investigations at Chaco Canyon Using Subsurface Radar' in T.R. Lyons (ed.) *Remote Sensing Experiments in Cultural Resource Studies* Chaco Center. USDI-NPS and the University of New Mexico. Albuquerque.

Vernon, R.W., McDonnell, G. and Schmidt, A. 1998 'The Geophysical Evaluation of an Iron-working Complex: Rievaulx and Environs, North Yorkshire' *Archaeological Prospection* 5(4) 181-202.

von Bandi, H.G. 1945 'Archaologische Erforschung des zukunftigen Stauseegebietes Rossens-Broc' *Schweizerischen Gesellschaft fur Urgeschichte* 30 100-6.

Walker, A.R. 1991 *Resistance Meter RM15 Maunal version 1.2.* Geoscan Research. Bradford. Unpublished.

Walker, A.R. 2000 'Multiplexed Resistivity Survey at the Roam Town of Wroxeter' *Archaeological Prospection* 7 119-32.

Waters, G.S. and Francis, P.D. 1958 'A nuclear magnetometer' *Journal of Scientific Instruments* 35 88-93.

Weston, D. 1996 'Soil Science and the Interpretation of Archaeological Sites: A Soil Survey and Magnetic Susceptibility Analysis of Altofts 'Henge, Normanton, West Yorkshire' *Archaeological Prospection* 3 39-50.

Weston, D. 2001 'Alluvium and Geophysical Prospection' *Archaeological Prospection* 8(4) 265-72.

Weymouth, J.W. 1986 'Archaeological site surveying program at the University of Nebraska' *Geophysics* 51 538-52.

White, R.H. and Barker, P.A. 1998 *Wroxeter: Life and Death of a Roman City* Tempus. Gloucestershire.

Wilson, D.R. 1984 'The Plan of Viroconium Cornoviorum' *Antiquity* 53 117-20.

INDEX

Entries in **bold** refer to figure numbers.